TAMING THE INFINITE

THE STORY OF MATHEMATICS

Ian Stewart is a world-renowned popularizer of mathematics, having won many awards for furthering the public understanding of science, including the Royal Society's Michael Faraday Medal and the Gold Medal of the Institute for Mathematics and Its Applications.

He is the author of over twenty popular science and mathematics titles, including *Why Beauty is Truth*, *Letters to a Young Mathematician*, *How to Cut a Cake*, *Does God Play Dice?*, *Nature's Numbers*, *Life's Other Secret* and *Flatterland*.

Professor Stewart is the mathematics consultant for *New Scientist*, wrote the 'Mathematical Recreations' column in Scientific American and has been a consultant for Encyclopaedia Britannica. He is an active research mathematician with over 170 published papers and is currently Professor of Mathematics at Warwick University. He was elected a Fellow of the Royal Society in 2001.

TAMING THE INFINITE

THE STORY OF MATHEMATICS

IAN STEWART

Quercus

First published in 2008 by Quercus

This paperback edition published 2009 by

Quercus Publishing Plc
21 Bloomsbury Square
London
WC1A 2NS

A catalogue record of this book is available from the British Library

ISBN 978 1 84724 768 1

All drawings by Tim Oliver

Printed and bound in Great Britain by Clays Ltd, St Ives plc

10 9 8 7 6 5

CONTENTS

PREFACE

Mathematics did not spring into being fully formed. It grew from the cumulative efforts of many people, from many cultures, who spoke many languages. Mathematical ideas that are still in use today go back more than 4000 years.

Many human discoveries are ephemeral – the design of chariot wheels was very important in New Kingdom Egypt, but it's not exactly cutting-edge technology today. Mathematics, in contrast, is often permanent. Once a mathematical discovery has been made, it becomes available for anyone to use, and thereby acquires a life of its own. Good mathematical ideas seldom go out of fashion, though their implementation can change dramatically. Methods for solving equations, discovered by the ancient Babylonians, are still in use today. We don't use their notation, but the historical link is undeniable.

In fact, most of the mathematics taught in schools is at least 200 years old. The advent of 'modern' mathematics curricula in the 1960s brought the subject into the 19th century. But contrary to appearances, mathematics has not stood still. Today, more new mathematics is created every week than the Babylonians managed in two thousand years.

The rise of human civilization and the rise of mathematics have gone hand in hand. Without Greek, Arab and Hindu discoveries in trigonometry, navigation across the open ocean would have been an even more hazardous task than it was when the great mariners opened up all six continents. Trade routes from China to Europe, or from Indonesia to the Americas, were held together by an invisible mathematical thread.

Today's society could not function without mathematics. Virtually everything that we now take for granted, from television to mobile phones, from wide-bodied passenger jets to satellite navigation systems in cars, from train schedules to medical scanners, relies on mathematical ideas and methods. Sometimes the mathematics is thousands of years old; sometimes it was discovered last week. Most of us never realize that it is present, working away behind the scenes to empower these miracles of modern technology.

This is unfortunate: it makes us think that technology works by magic, and leads us to expect new miracles on a daily basis. On the other hand, it is also entirely natural: we want to use these miracles as easily and with as little thought as possible. The user should not be burdened by unnecessary information about the underlying trickery that makes the miracles possible. If every airline passenger were made to pass an examination in trigonometry before boarding the aircraft, few of us would ever leave the ground. And while that might reduce our carbon footprint, it would also make our world very small and parochial.

To write a truly comprehensive history of mathematics is virtually impossible. The subject is now so broad, so complicated and so technical, that even an expert would find such a book unreadable – leaving aside the impossibility of anyone writing it. Morris Kline came close with his epic *Mathematical Thought from Ancient to Modern Times*. It is more than 1200 pages long, in small print, and it misses out almost everything that has happened in the last 100 years.

This book is much shorter, which means that I have had to be selective, especially when it comes to mathematics of the 20th and 21st centuries. I am acutely conscious of all the important topics that I have had to omit. There is no algebraic geometry, no cohomology theory, no finite element analysis, no wavelets. The list of what's missing is far longer than the list of what is included. My choices have been guided by what background knowledge readers are likely to possess, and what new ideas can be explained succinctly.

The story is roughly chronological within each chapter, but the chapters are organized by topic. This is necessary to have any kind of coherent narrative; if I put everything into chronological order the discussion would flit randomly from one topic to another, without any sense of direction. This might be closer to the actual history, but it would make the book unreadable. So each new chapter begins by stepping back into the past, and then touching upon some of the historical milestones that were passed as the subject developed. Early chapters stop a long way in the past; later chapters sometimes make it all the way to the present.

I have tried to give a flavour of modern mathematics, by which I mean anything done in the last 100 years or so, by selecting topics that readers may have heard of, and relating them to the overall historical trends. The omission of a topic does not imply that it lacks importance, but I think it makes more sense to spend a few pages talking about Andrew Wiles's proof of Fermat's Last Theorem – which most readers will have heard of – than, say, non-commutative geometry, where the background alone would occupy several chapters.

In short, this is *a* history, not *the* history. And it is history in the sense that it tells a story about the past. It is not aimed at professional historians, it does not make the fine distinctions that they find necessary, and it often describes the ideas of the past through the eyes of the present. The latter is a cardinal sin for a historian, because it makes it look as though the ancients were somehow striving towards our current way of thinking. But I think it is both defensible and essential if the main objective is to start from what we now know and ask where those ideas came from. The Greeks did not study the ellipse in order to make possible Kepler's theory of planetary orbits, and Kepler did not formulate his three laws of planetary motion in order for Newton to turn them into his law of gravity. However, the story of Newton's law rests heavily on Greek work on the ellipse and Kepler's analysis of observational data.

A sub-theme of the book is the practical uses of mathematics. Here I have provided a very eclectic sample of applications, both past and present. Again, the omission of any topic does not indicate that it lacks importance.

Mathematics has a long, glorious, but somewhat neglected history, and the subject's influence on the development of human culture has been immense. If this book conveys a tiny part of that story, it will have achieved what I set out to do.

Coventry, May 2007

CHAPTER 1

Tokens, Tallies and Tablets

The birth of numbers

Mathematics began with numbers, and numbers are still fundamental, even though the subject is no longer limited to numerical calculations. By building more sophisticated concepts on the basis of numbers, mathematics has developed into a broad and varied area of human thought, going far beyond anything that we encounter in a typical school syllabus. Today's mathematics is more about structure, pattern and form than it is about numbers as such. Its methods are very general, and often abstract. Its applications encompass science, industry, commerce – even the arts. Mathematics is universal and ubiquitous.

It started with numbers

Over many thousands of years, mathematicians from many different cultures have created a vast superstructure on the foundations of number: geometry, calculus, dynamics, probability, topology, chaos, complexity and so on. The journal *Mathematical Reviews*, which keeps track of every new mathematical publication, classifies the subject into nearly a hundred major areas, subdivided into several thousand specialities. There are more than 50,000 research mathematicians in the world, and they publish more than a million pages of new mathematics every year. Genuinely new mathematics, that is, not just small variations on existing results.

Mathematicians have also burrowed into the logical foundations of their subject, discovering concepts even more fundamental than numbers – mathematical logic, set theory. But again the main motivation, the starting point from which all else flows, is the concept of number.

Numbers seem very simple and straightforward, but appearances are deceptive. Calculations with numbers can be hard; getting the right number can be difficult. And even then, it is much easier to use numbers than to specify what they really are. Numbers count things, but they are not things, because you can pick up two cups, but you can't pick up the number 'two'. Numbers are denoted by symbols, but different cultures use different symbols for the same number. Numbers are abstract, yet we base our society on them and it would not function without them. Numbers are some kind of mental construct, yet we feel that they would continue to have meaning even if humanity were wiped out by a global catastrophe and there were no minds left to contemplate them.

Writing numbers

The history of mathematics begins with the invention of written symbols to denote numbers. Our familiar system of digits 0, 1, 2, 3, 4, 5, 6, 7, 8, 9 to represent all conceivable numbers, however large, is a relatively new invention; it came into being about 1500 years ago, and its extension to decimals, which lets us represent numbers to high precision, is no more than 450 years old. Computers, which have embedded mathematical calculations so deeply into our culture that we no longer notice their presence, have been with us for a mere 50 years; computers powerful enough and fast enough to be useful in our homes and offices first became widespread about 20 years ago.

Without numbers, civilization as we now know it could not exist. Numbers are everywhere, hidden servants that scurry around behind the scenes – carrying messages, correcting our spelling when we type, scheduling our holiday flights to the Caribbean, keeping track of our goods, ensuring that our medicines are safe and effective. And, for balance, making nuclear weapons possible, and guiding bombs and missiles to their targets. Not every application of mathematics has improved the human condition.

How did this truly enormous numerical industry arise? It all began with little clay tokens, 10,000 years ago in the Near East.

Even in those days, accountants were keeping track of who owned what, and how much – even though writing had not then been invented, and there were no symbols for numbers. In place of number symbols, those ancient accountants used small clay tokens. Some were cones, some were spheres and some were shaped like eggs. There were cylinders, discs and pyramids. The archaeologist Denise Schmandt-Besserat deduced that these tokens represented basic staples of the time. Clay spheres represented bushels of grain, cylinders stood for animals, eggs for jars of oil. The earliest tokens date back to 8000 BC, and they were in common use for 5000 years.

As time passed, the tokens became more elaborate and more specialized. There were decorated cones to represent loaves of bread, and diamond-shaped slabs to represent beer. Schmandt-Besserat realized that these tokens were much more than an accountancy device. They were a vital first step along the path to number symbols, arithmetic and mathematics. But that initial step was rather strange, and it seems to have happened by accident.

It came about because the tokens were used to keep records, perhaps for tax purposes or financial ones, or as legal proof of ownership. The advantage of tokens was that the accountants could quickly arrange them in patterns, to work out how many animals or how much grain someone owned or owed. The disadvantage was that tokens could be counterfeited. So to make sure that no one interfered with the accounts, the accountants wrapped the tokens in clay envelopes – in effect, a kind of seal. They could quickly find out how many tokens were inside any given envelope, and of what kind, by breaking the envelope open. They could always make a new envelope for later storage.

However, repeatedly breaking open an envelope and renewing it was a rather inefficient way to find out what was inside, and the bureaucrats of ancient Mesopotamia thought of something better. They inscribed symbols on the envelope, listing the tokens that it contained. If there were seven spheres inside, the accountants would draw seven pictures of spheres in the wet clay of the envelope.

At some point the Mesopotamian bureaucrats realized that once they had drawn the symbols on the outside of the envelope, they didn't actually need the contents, so they didn't need to break open the envelope to find out which tokens were inside it. This obvious but crucial step effectively created a set of written number symbols, with different shapes for different classes of goods. All other number symbols, including those we use today, are the intellectual descendants of this ancient bureaucratic device. In fact, the replacement of tokens by symbols may also have constituted the birth of writing itself.

Tally marks

These clay marks were by no means the earliest examples of number-writing, but all earlier instances are little more than scratches, tally marks, recording numbers as a series of strokes – such as ||||||||||||| to represent the number 13. The oldest known marks of this kind – 29 notches carved on a baboon's leg bone – are about 37,000 years old. The bone was found in a cave in the Lebombo mountains, on the border between Swaziland and

Tally marks have the advantage that they can be built up one at a time, over long periods, without altering or erasing previous marks. They are still in use today, often in groups of five, with the fifth stroke cutting diagonally across the previous four.

The presence of tally marks can still be seen in modern numerals. Our symbols 1, 2, 3 are derived from a single stroke, two horizontal strokes linked by a sloping line, and three horizontal strokes linked by a sloping line.

South Africa, so the cave is known as the Border Cave, and the bone is the Lebombo bone. In the absence of a time machine, there is no way to be certain what the marks represented, but we can make informed guesses. A lunar month contains 28 days, so the notches may have related to the phases of the Moon.

There are similar relics from ancient Europe. A wolf bone found in former Czechoslovakia has 57 marks arranged in eleven groups of five with two left over, and is about 30,000 years old. Twice 28 is 56, so this might perhaps be a two-month long lunar record. Again, there seems to be no way to test this suggestion. But the marks look deliberate, and they must have been made for some reason.

Another ancient mathematical inscription, the Ishango bone from Zaire, is 25,000 years old (previous estimates of 6000–9000 years were revised in 1995). At first sight the marks along the edge of the bone seem almost random, but there may be hidden patterns. One row contains the prime numbers between 10 and 20, namely 11, 13, 17 and 19, whose sum is 60. Another row contains 9, 11, 19 and 21, which also sum to 60. The third row resembles a method sometimes used to multiply two numbers together by repeated

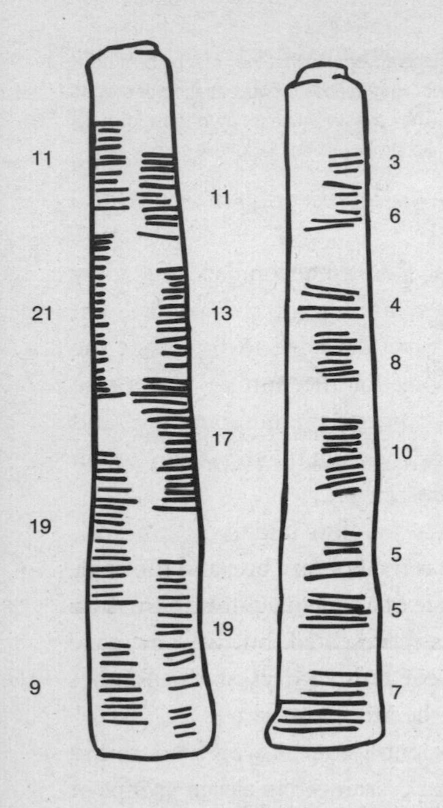

The Ishango bone showing the patterns of marks and the numbers they may represent.

doubling and halving. However, the apparent patterns may just be coincidental, and it has also been suggested that the Ishango bone is a lunar calendar.

The first numerals

The historical path from accountants' tokens to modern numerals is long and indirect. As the millennia passed, the people of Mesopotamia developed agriculture, and their nomadic way of life gave way to permanent settlements, in a series of city states – Babylon, Eridu, Lagash, Sumer, Ur. The early symbols inscribed on tablets of wet clay mutated into pictographs – symbols that represent words by simplified pictures of what the words mean – and the pictographs were further simplified by assembling them from a small number of wedge-shaped marks, impressed into the clay using a dried reed with a flat, sharpened end. Different types of wedge could be made by holding the reed in different ways. By 3000 BC, the Sumerians had developed an elaborate form of writing, now called *cuneiform* or wedge-shaped.

The history of this period is complicated, with different cities becoming dominant at different times. In particular, the city of Babylon came to prominence, and about a million Babylonian clay tablets have been dug from the Mesopotamian sands. A few hundred of these deal with mathematics and astronomy, and they show that the Babylonian knowledge of both subjects was extensive. In particular, the Babylonians were accomplished astronomers, and they developed a systematic and sophisticated symbolism for numbers, capable of representing astronomical data to high precision.

The Babylonian number symbols go well beyond a simple tally system, and are the earliest known symbols to do so. Two different types of wedge are used: a thin vertical wedge to represent the number 1, and a fat horizontal one for the number 10. These wedges are arranged in groups to indicate the numbers 2–9 and 20–50. However, this pattern stops at 59, and the thin wedge then takes on a second meaning, the number 60.

The Babylonian number system is therefore said to be 'base 60', or sexagesimal. That is, the value of a symbol may be some number, or 60 times such a number, or 60 times 60 times such a

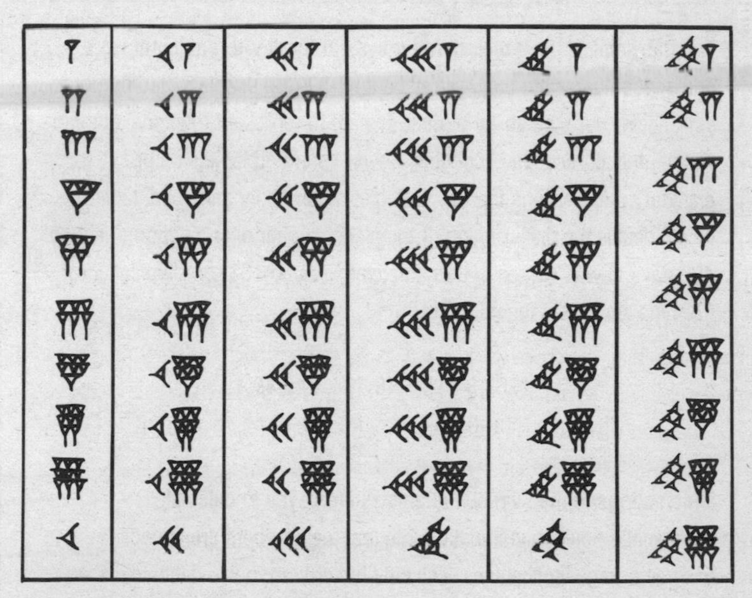

Babylonian symbols for the numbers 1–59

number, depending on the symbol's position. This is similar to our familiar decimal system, in which the value of a symbol is multiplied by 10, or by 100, or by 1000, depending on its position. In the number 777, for instance, the first 7 means 'seven hundred', the second means 'seventy' and the third means 'seven'. To a Babylonian, a series of three repetitions [cuneiform symbols] of the symbol for '7' would have a different meaning, though based on a similar principle. The first symbol would mean 7 × 60 × 60, or 25,200; the second would mean 7 × 60 = 420; the third would mean 7. So the group of three symbols would mean 25,200 + 420 + 7, which is 25,627 in our notation. Relics of Babylonian base-60 numbers can still be found today. The 60 seconds in a minute, 60 minutes in an hour and 360 degrees in a full circle, all trace back to ancient Babylon.

What numbers did for them

The Babylonians used their number system for day-to-day commerce and accountancy, but they also used it for a more sophisticated purpose: astronomy. Here the ability of their system to represent fractional numbers with high accuracy was essential. Several hundred tablets record planetary data. Among them is a single, rather badly damaged, tablet which details the daily motion of the planet Jupiter over a period of about 400 days. It was written in Babylon itself, around 163 BC. A typical entry from the tablet lists the numbers

126 8 16;6,46,58 -0;0,45,18
-0;0,11,42 +0;0,0,10,

which correspond to various quantities employed to calculate the planet's position in the sky. Note that the numbers are stated to three sexagesimal places – slightly better than to five decimal places.

Because it is awkward to typeset cuneiform, scholars write Babylonian numerals using a mixture of our base-10 notation and their base-60 notation. So the three repetitions of the cuneiform symbol for 7 would be written as 7,7,7. And something like 23,11,14 would indicate the Babylonian symbols for 23, 11 and 14 written in order, with the numerical value $(23 \times 60 \times 60) + (11 \times 60) + 14$, which comes to 83,474 in our notation.

Symbols for small numbers

Not only do we use ten symbols to represent arbitrarily large numbers: we also use the same symbols to represent arbitrarily small ones. To do this, we employ the decimal point, a small dot. Digits to the left of the dot represent whole numbers; those to the right of the dot represent fractions. Decimal fractions are multiples of one tenth, one hundredth, and so on. So 25.47, say, means 2 tens plus 5 units plus 4 tenths plus 7 hundredths.

The Babylonians knew this trick, and they used it to good effect in their astronomical observations. Scholars denote the Babylonian equivalent of the decimal point by a semicolon (;), but this is a sexagesimal point and the numbers to its right are multiples of $1/60$, $(1/60 \times 1/60) = 1/3600$, and so on. As an example, the list of numbers 12,59;57,17 means

$$12 \times 60 + 59 + 57/60 + 17/3600$$

which is roughly 779.955.

Nearly 2000 Babylonian tablets with astronomical information are known, though many of these are fairly routine, consisting of descriptions of ways to predict eclipses, tables of regular astronomical events and shorter extracts. About 300 tablets are more ambitious and more exciting; they tabulate observations of the motion of Mercury, Mars, Jupiter and Saturn, for example.

Fascinating as it may be, Babylonian astronomy is somewhat tangential to our main story, which is Babylonian pure mathematics. But it seems likely that the application to astronomy was a spur to the pursuit of the more cerebral areas of that subject. So it is a good idea to recognize just how accurate the Babylonian astronomers were when it came to observing heavenly events. For example, they found the orbital period of Mars (strictly, the time between successive appearances in the same position in the sky) to be 12,59;57,17 days in their notation – roughly 779.955 days, as noted above. The modern figure is 779.936 days.

The Ancient Egyptians

Perhaps the greatest of the ancient civilizations was that of Egypt, which flourished on the banks of the Nile and in the Nile Delta between 3150 BC and 31 BC, with an extensive earlier 'pre-dynastic' period stretching back to 6000 BC, and a gradual tailing off under the Romans from 31 BC onwards. The Egyptians were accomplished builders, with a highly developed system of religious beliefs and ceremonies, and they were obsessive record-keepers. But their mathematical achievements were modest compared to the heights scaled by the Babylonians.

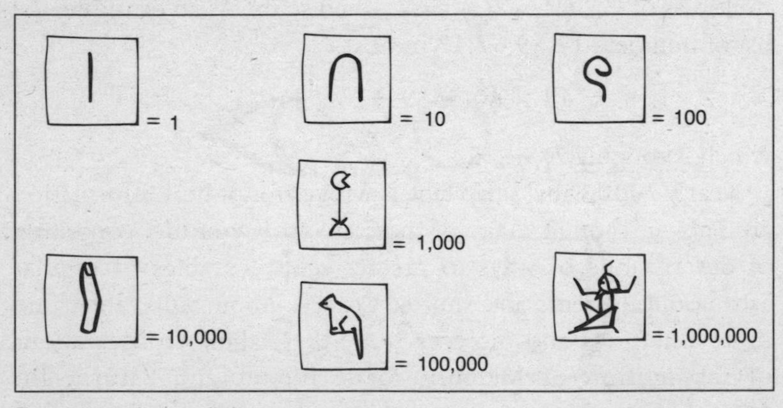

Egyptian number symbols

The ancient Egyptian system for writing whole numbers is simple and straightforward. There are symbols for the numbers 1, 10, 100, 1000 and so on. Repeating these symbols up to nine times, and then combining the results, can represent any whole number. For example, to write the number 5724, the Egyptians would group together five of their symbols for 1000, seven symbols for 100, two symbols for 10 and four symbols for 1.

The number 5724 in Egyptian hieroglyphs

Fractions caused the Egyptians severe headaches. At various periods, they used several different notations for fractions. In the Old Kingdom (2700–2200 BC), a special notation for our fractions, $\frac{1}{2}$, $\frac{1}{4}$, $\frac{1}{8}$, $\frac{1}{16}$, $\frac{1}{32}$ and $\frac{1}{64}$, was obtained by repeated halving. These symbols used parts of the 'eye of Horus' or 'wadjet-eye' hieroglyph.

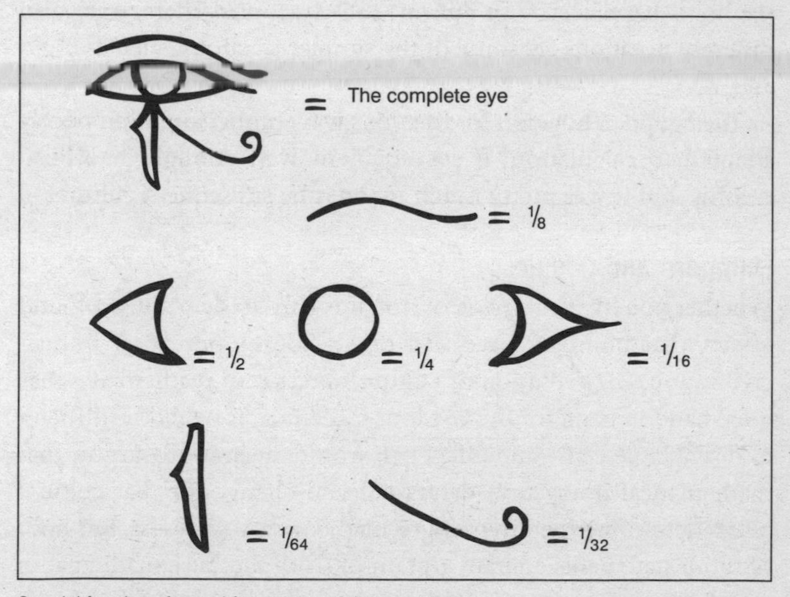

Special fractions formed from parts of the wadjet eye

The best known Egyptian system for fractions was devised during the Middle Kingdom (2000–1700 BC). It starts with a notation for any fraction of the form $1/n$, where n is a positive integer. The symbol ⬭ (the hieroglyph for the letter R) is written over the top of the standard Egyptian symbols for n. So, for example, $^1/_{11}$ is written . Other fractions are then expressed by adding together several of these 'unit fractions'. For instance, $^5/_6 = {}^1/_2 + {}^1/_3$.

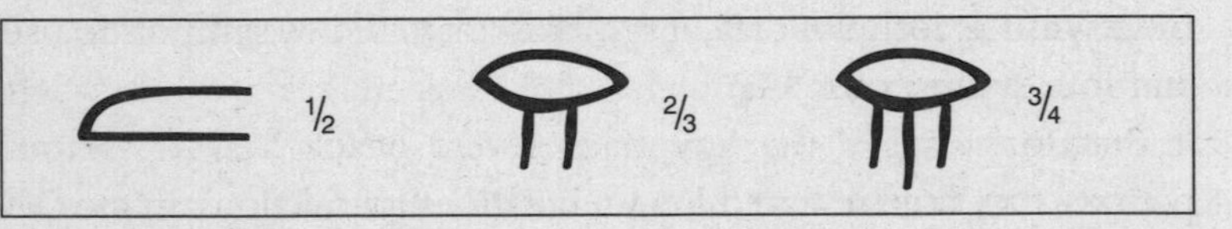

Special symbols for special fractions

Interestingly, the Egyptians did not write $^2/_5$ as $^1/_5 + {}^1/_5$. Their rule seems to have been: use *different* unit fractions. There were also different notations for some of the simpler fractions, such as $^1/_2$, $^2/_3$ and $^3/_4$.

The Egyptian notation for fractions was cumbersome and poorly adapted to calculation. It served them well enough in official records, but it was pretty much ignored by subsequent cultures.

Numbers and people

Whether you like arithmetic or not, it is hard to deny the profound effects that numbers have had on the development of human civilization. The evolution of culture, and that of mathematics, has gone hand in hand for the last four millennia. It would be difficult to disentangle cause and effect – I would hesitate to argue that mathematical innovation drives cultural change, or that cultural needs determine the direction of mathematical progress. But both of those statements contain a grain of truth, because mathematics and culture co-evolve.

There is a significant difference, though. Many cultural changes are clearly apparent. New kinds of housing, new forms of transport, even new ways to organize government bureaucracies, are relatively obvious to every citizen. Mathematics, however, mostly takes place behind the scenes. When the Babylonians used their astronomical observations to predict solar eclipses, for instance, the average citizen was impressed by how accurately the priests forecast this astonishing event, but even the majority of the priesthood had little or no idea of the methods employed. They knew how to read tablets listing eclipse data, but what mattered was how to use them. How they had been constructed was an arcane art, best left to specialists.

Some priests may have had good mathematical educations – all trained scribes did, and trainee priests took much the same lessons as scribes, in their early years – but an appreciation of mathematics wasn't really necessary to *enjoy* the benefits that flowed from new discoveries in that subject. So it has ever been, and no doubt always will be. Mathematicians seldom get credit for changing our world. How many times do you see all kinds of modern miracles credited to computers, without the slightest appreciation that computers only work effectively if they are programmed to use sophisticated algorithms, that is procedures to solve problems, and that the basis of almost all algorithms is mathematics?

The main mathematics that does lie on the surface is arithmetic. And the invention of pocket calculators, tills that tot up how much you have to pay and tax accountants who do the sums for you, for a fee, are pushing even arithmetic further behind the scenes. But at least most of us are aware that the arithmetic is there. We are wholly dependent on numbers, be it for keeping track of legal obligations, levying taxes, communicating instantly with the far side of the planet, exploring the surface of Mars or assessing the latest wonder drug. All of these things trace back to ancient Babylon, and to the scribes and teachers who discovered effective ways to record

What numbers do for us

Most upmarket modern cars now come equipped with satnav – satellite navigation. Stand-alone satnav systems can be purchased relatively cheaply. A small device, affixed to your car, then tells you exactly where you are at any moment and displays a map – often in fancy colour graphics and perspective – showing the neighbouring roads. A voice system can even tell you where to go to reach a specified destination. If this sounds like something out of science fiction, it is. An essential component, not part of the small box attached to the car, is the Global Positioning System (GPS), which comprises 24 satellites orbiting the Earth, sometimes more as replacements are launched. These satellites send out signals, and these signals can be used to deduce the location of the car to within a few metres.

Mathematics comes into play in many aspects of the GPS network, but here we mention just one: how the signals are used to work out the location of the car.

Radio signals travel at the speed of light, which is roughly 300,000 kilometres per second. A computer on board the car – a chip in the box you buy – can work out the distance from your car to any given satellite if it knows how long the signal has taken to travel from the satellite to your car. This is typically about one tenth of a second, but precise time measurement is now easy. The trick is to structure the signal so that it contains information about timing.

In effect, the satellite and the receiver in the car both play the same tune, and compare its timing. The 'notes' coming from the satellite will lag slightly behind those produced in the car. In this analogy, the tunes might go like this:

CAR	... feet, in ancient times, walk upon England's ...
SATELLITE	... And did those feet, in ancient times, walk ...

Here the satellite's song is lagging some three words behind the same song in the car. Both satellite and receiver must generate the same 'song', and successive 'notes' must be distinctive, so that the timing difference is easy to observe.

Of course, the satnav system doesn't actually use a song. The signal is a series of brief pulses whose duration is determined by a 'pseudo-random code'. This is a series of numbers, which looks random but is actually based on some mathematical rule. Both the satellite and the receiver know the rule, so they can generate the same sequence of pulses.

numbers and calculate with them. They used their arithmetical skills for two main purposes: down-to-earth everyday affairs of ordinary human beings, such as land-measurement and accountancy, and highbrow activities like predicting eclipses or recording the movements of planets across the night-time sky.

We do the same today. We use simple mathematics, little more than arithmetic, for hundreds of tiny tasks – how much anti-parasite treatment to put into a garden fishpond, how many rolls of wallpaper to buy to paper the bedroom, whether it will save money to travel further for cheaper petrol. And our culture uses sophisticated mathematics for science, technology and increasingly for commerce too. The inventions of number notation and arithmetic rank alongside those of language and writing as some of the innovations that differentiate us from trainable apes.

CHAPTER 2

The Logic of Shape

First steps in geometry

There are two main types of reasoning in mathematics: symbolic and visual. Symbolic reasoning originated in number notation, and we will shortly see how it led to the invention of algebra, in which symbols can represent general numbers ('the unknown') rather than specific ones ('7'). From the Middle Ages onwards, mathematics came to rely increasingly heavily on the use of symbols, as a glance at any modern mathematics text will confirm.

The beginnings of geometry

As well as symbols, mathematicians use diagrams, opening up various types of visual reasoning. Pictures are less formal than symbols, and their use has sometimes been frowned upon for that reason. There is a widespread feeling that a picture is somehow less rigorous, logically speaking, than a symbolic calculation. It is true that pictures leave more room for differences of interpretation than symbols. Additionally, pictures can contain hidden assumptions – we cannot draw a 'general' triangle; any triangle we draw has a particular size and shape, which may not be representative of an arbitrary triangle. Nonetheless, visual intuition is such a powerful feature of the human brain that pictures play a prominent role in mathematics. In fact, they introduce a second major concept into the subject, after number. Namely, shape.

Mathematicians' fascination with shapes goes back a long way. There are diagrams on Babylonian tablets. For example, the tablet catalogued as YBC 7289 shows a square and two diagonals. The sides of the square are marked with cuneiform numerals for 30. Above one diagonal is marked 1;24,51,10, and below it 42;25,35, which is its product by 30 and therefore the length of that diagonal. So 1;24,51,10 is the length of the diagonal of a smaller square, with sides 1 unit. Pythagoras's Theorem tells us that this diagonal is the square root of two, which we write as $\sqrt{2}$. The approximation 1;24,51,10 to $\sqrt{2}$ is very good, correct to 6 decimal places.

The first systematic use of diagrams, together with a limited use of symbols and a heavy dose of logic, occurs in the geometric writings of Euclid of Alexandria. Euclid's work followed a tradition that went back at least to the Pythagorean cult, which flourished around 500 BC, but Euclid insisted that any mathematical statement must be given a logical proof before it could be assumed to be true. So Euclid's writings combine two distinct innovations: the use of pictures and the logical structure of proofs. For centuries, the word 'geometry' was closely associated with both.

In this chapter we follow the story of geometry from Pythagoras, through Euclid and his forerunner Eudoxus, to the late period of classical Greece and Euclid's successors Archimedes and Apollonius. These early geometers paved the way for all later work on visual thinking in mathematics. They also set standards of logical proof that were not surpassed for millennia.

Pythagoras

Today we almost take it for granted that mathematics provides a key to the underlying laws of nature. The first recorded systematic thinking along those lines comes from the Pythagoreans, a rather mystical cult dating from around 500 BC. Its founder, Pythagoras, was born on Samos around 569 BC. When and where he died is a mystery, but in 460 BC the cult that he founded was attacked and destroyed, its meeting places wrecked and burned. In one, the house of Milo in Croton, more than 50 Pythagoreans were slaughtered. Many survivors fled to Thebes in Upper Egypt. Possibly Pythagoras was one of them, but even this is conjectural, for legends aside, we know virtually nothing about Pythagoras. His name is well known, mainly because of his celebrated theorem about right-angled triangles, but we don't even know whether Pythagoras proved it.

We know much more about the Pythagoreans' philosophy and beliefs. They understood that mathematics is about abstract concepts, not reality. However, they also believed that these abstractions were somehow embodied in 'ideal' concepts, existing in some strange realm of the imagination, so that, for instance, a circle drawn in sand with a stick is a flawed attempt to be an ideal circle, perfectly round and infinitely thin.

The most influential aspect of the Pythagorean cult's philosophy is the belief that the universe is founded on numbers. They expressed this belief in mythological symbolism, and supported it with empirical observations. On the mystic side, they considered the number 1 to be the prime source of everything in the universe. The numbers 2 and 3 symbolized the female and male principles. The number 4 symbolized harmony, and also the four elements (earth, air, fire, water) out of which everything is made. The Pythagoreans believed that the number 10 had deep mystical significance, because 10 = 1 + 2 + 3 + 4, combining prime unity, the female principle, the male principle and the four elements. Moreover, these numbers formed a triangle, and the whole of Greek geometry hinged upon properties of triangles.

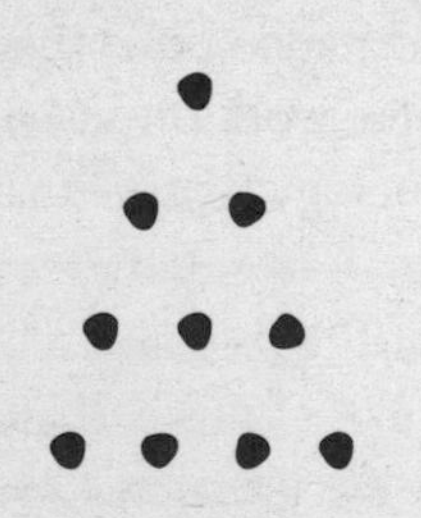

The number ten forms a triangle

Harmony of the World

The main empirical support for the Pythagorean concept of a numerical universe came from music, where they had noticed some remarkable connections between harmonious sounds and simple numerical ratios. Using simple experiments, they discovered that if a plucked string produces a note with a particular pitch, then a string half as long produces an extremely harmonious note, now called the octave. A string two-thirds as long produces the next most harmonious note, and one three-quarters as long also produces a harmonious note.

Today these numerical aspects of music are traced to the physics of vibrating strings, which move in patterns of waves. The number of waves that can fit into a given length of string is a whole number, and these whole numbers determine the simple numerical ratios. If the numbers do not form a simple ratio then the corresponding notes interfere with each other, forming discordant 'beats' which are unpleasant to the ear. The full story is more complex, involving what the brain becomes accustomed to, but there is a definite physical rationale behind the Pythagorean discovery.

The Pythagoreans recognized the existence of nine heavenly bodies, Sun, Moon, Mercury, Venus, Earth, Mars, Jupiter and Saturn, plus the Central Fire, which differed from the Sun. So important was the number 10 in their view of cosmology that they believed there was a tenth body, Counter-Earth, perpetually hidden from us by the Sun.

As we have seen, the whole numbers 1, 2, 3, ..., naturally lead to a second type of number, fractions, which mathematicians call *rational numbers*. A rational number is a fraction a/b where a, b are whole numbers (and b is non-zero, otherwise the fraction makes no sense). Fractions subdivide whole numbers into arbitrarily fine parts, so that in particular the length of a line in a geometric figure can be approximated as closely as we wish by a rational number. It seems natural to imagine that enough subdivision would hit the number exactly; if so, all lengths would be rational.

If this were true, it would make geometry much simpler, because any two lengths would be whole number multiples of a common (perhaps small) length, and so could be obtained by fitting lots of copies of this common length together. This may not sound very important, but it would make the whole theory of lengths, areas and especially similar figures – figures with the same shape but different sizes – much simpler. Everything could be proved using diagrams formed from lots and lots of copies of one basic shape.

Unfortunately, this dream cannot be realized. According to legend, one of the followers of Pythagoras, Hippasus of Metapontum, discovered that this statement was false. Specifically, he proved that the diagonal of a unit square (a square with sides one unit long) is irrational: not an exact fraction. It is said (on dubious grounds, but it's a good story) that he made the mistake of announcing this fact when the Pythagoreans were crossing the Mediterranean by boat, and his fellow cult-members were so incensed that they threw him overboard and he drowned. More likely he was just expelled from the cult. Whatever his punishment, it seems that the Pythagoreans were not pleased by his discovery.

The modern interpretation of Hippasus's observation is that $\sqrt{2}$ is irrational. To the Pythagoreans, this brutal fact was a body-blow to their almost religious belief that the universe was rooted in numbers – by which they meant whole numbers. Fractions – ratios of whole numbers – fitted neatly enough into this world-view, but numbers that were provably not fractions did not. And so, whether drowned or expelled, poor Hippasus became one of the early victims of the irrationality, so to speak, of religious belief.

Taming irrationals

Eventually, the Greeks found a way to handle irrationals. It works because any irrational number can be approximated by a rational number. The better the approximation, the more complicated that rational becomes, and there is always some error. But by making the error smaller and smaller, there is a prospect of approaching the properties of irrationals by exploiting analogous properties of approximating rational numbers. The problem is to set this idea up in a way that is compatible with the Greek approach to geometry and proof. This turns out to be feasible, but complicated.

The Greek theory of irrationals was invented by Eudoxus around 370 BC. His idea is to represent any magnitude, rational or irrational, as the ratio of two lengths – that is, in terms of a pair of lengths. Thus two-thirds is represented by two lines, one of length two and

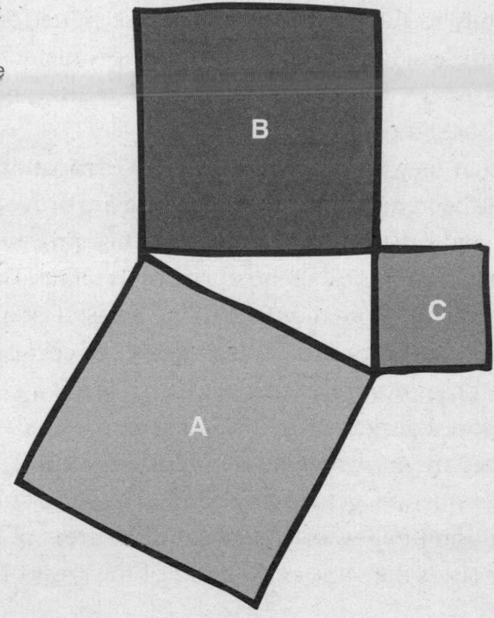

Pythagoras's Theorem: if the triangle has a right angle, then the largest square, A, has the same area as the other two, B and C, combined

one of length three (a ratio 2:3). Similarly, $\sqrt{2}$ is represented by the pair formed by the diagonal of a unit square, and its side (a ratio $\sqrt{2}$:1). Note that both pairs of lines can be constructed geometrically.

The key point is to define when two such ratios are *equal*. When is $a{:}b = c{:}d$? Lacking a suitable number system, the Greeks could not do this by dividing one length by the other and comparing $a \div b$ with $c \div d$. Instead, Eudoxus found a cumbersome but precise method of comparison that could be performed within the conventions of Greek geometry. The idea is to try to compare a and c by forming *integer* multiples ma and nc. This can be done by joining m copies of a end to end, and similarly n copies of c. Use the same two multiples m and n to compare mb and nd. If the ratios $a{:}b$ and $c{:}d$ are not equal, says Eudoxus, then we can find m and n to exaggerate the difference, to such an extent that $ma > nc$ but $mb < nd$. Indeed, we can *define* equality of ratios that way.

This definition takes some getting used to. It is tailored very carefully to the limited operations permitted in Greek geometry. Nonetheless, it works; it let the Greek geometers take theorems that could easily be proved for rational ratios, and extend them to irrational ratios.

Often they used a method called 'exhaustion', which let them prove theorems that we would nowadays prove using the idea of a limit and calculus. In this manner they proved that the area of a circle is proportional to the square of its radius. The proof starts from a simpler fact, found in Euclid: the areas of two similar *polygons* are in the same proportion as the squares of corresponding sides. The circle poses new problems because it is not a polygon. The Greeks therefore considered two sequences of regular polygons whose vertices are on the circle: one inside the circle, the other outside. Both sequences get closer and closer to the circle, and Eudoxus's definition implies that the ratio of the areas of the approximating polygons is the same as the ratio of the areas of the circles.

Euclid

The best-known Greek geometer, though probably not the most original mathematician, is Euclid of Alexandria. Euclid was a great synthesizer, and his geometry text, the *Elements*, became an all-time bestseller. Euclid wrote at least ten texts on mathematics, but only five of them survive – all through later copies, and then only in part. We have no original documents from ancient Greece. The five Euclidean survivors are the *Elements*, the *Division of Figures*, the *Data*, the *Phaenomena* and the *Optics*.

The *Elements* is Euclid's geometrical masterpiece, and it provides a definitive treatment of the geometry of two dimensions (the plane) and three dimensions (space). The *Division of Figures* and the *Data* contain various supplements and comments on geometry. The *Phaenomena* is aimed at astronomers, and deals with spherical geometry, the geometry of figures drawn on the surface of a sphere. The *Optics* is also geometric, and might best be thought of as an early investigation of the geometry of perspective – how the human eye transforms a three-dimensional scene into a two-dimensional image.

Perhaps the best way to think of Euclid's work is as an examination of the logic of spatial relationships. If a shape has certain properties, these may logically imply other properties. For example, if a triangle has all three sides equal – an equilateral triangle – then all three angles must be equal. This type of statement, listing some assumptions and then stating their logical consequences, is called a theorem. This particular theorem relates a property of the sides of a triangle to a property of its angles. A less intuitive and more famous example is Pythagoras's Theorem.

The *Elements* breaks up into 13 separate books, which follow each other in logical sequence. They discuss the geometry of the plane, and some aspects of the geometry of space. The climax is the proof that there are precisely five regular solids: the tetrahedron, cube, octahedron, dodecahedron and icosahedron. The basic shapes

Regular Polyhedra

A solid is regular (or Platonic) if it is formed from identical faces, arranged in the same way at each vertex, with each face a regular polygon. The Pythagoreans knew of five such solids:

The five Platonic solids

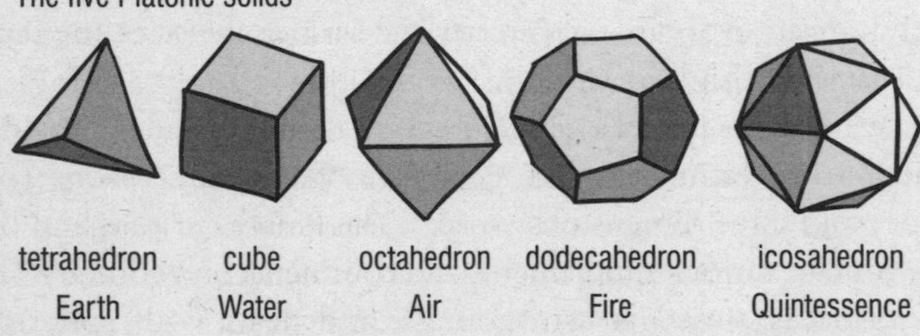

- The tetrahedron, formed from four equilateral triangles.
- The cube (or hexahedron), formed from six squares.
- The octahedron, formed from eight equilateral triangles.
- The dodecahedron, formed from 12 regular pentagons.
- The icosahedron, formed from 20 equilateral triangles.

They associated them with the four elements of antiquity – earth, air, fire and water – and with a fifth element, quintessence, which means fifth element.

permitted in plane geometry are straight lines and circles, often in combination – for instance, a triangle is formed from three straight lines. In spatial geometry we also find planes, cylinders and spheres.

To modern mathematicians, what is most interesting about Euclid's geometry is not its content, but its logical structure. Unlike his predecessors, Euclid does not merely assert that some theorem is true. He provides a proof.

What is a proof? It is a kind of mathematical story, in which each step is a logical consequence of some of the previous steps. Every statement that is asserted has to be justified by referring it back to previous statements and showing that it is a logical consequence of them. Euclid realized that this process cannot go back indefinitely: it has to start somewhere, and those initial statements cannot themselves be proved – or else the process of proof actually starts somewhere different.

To start the ball rolling, Euclid started by listing a number of definitions: clear, precise statements of what certain technical terms, such as *line* or *circle*, mean. A typical definition is 'an obtuse angle is an angle greater than a right angle'. The definitions gave him the terminology that he needed to state his unproved assumptions, which he classified into two types: *common notions* and *postulates*. A typical common notion is 'things which are equal to the same thing are equal to one another'. A typical postulate is 'all right angles are equal to one another'.

Nowadays we would lump both types together and call them axioms. The axioms of a mathematical system are the underlying assumptions that we make about it. We think of the axioms as the rules of the game, and insist that the game is played according to the rules. We no longer ask whether the rules are true – we no longer think that only one game can be played. Anyone who wants to play that particular game must accept the rules; if they don't, they are free to play a different game, but it won't be the one determined by those particular rules.

In Euclid's day, and for nearly 2000 years afterwards, mathematicians didn't think that way at all. They generally viewed the axioms as self-evident truths, so obvious that no one could seriously question them. So Euclid did his best to make all of his axioms obvious – and very nearly succeeded. But one axiom, the 'parallel axiom', is unusually complicated and unintuitive, and many people tried to deduce it from simpler assumptions. Later, we'll see the remarkable discoveries that this led to.

From these simple beginnings, the *Elements* proceeded, step by

Euclid of Alexandria

325–265 BC

Euclid is famous for his geometry book the *Elements*, which was a prominent – indeed, the leading – text in mathematical teaching for two millennia.

We know very little about Euclid's life. He taught at Alexandria. Around 45 BC the Greek philosopher Proclus wrote:

'Euclid ... lived in the time of the first Ptolemy, for Archimedes, who followed closely upon the first Ptolemy, makes mention of Euclid ... Ptolemy once asked [Euclid] if there were a shorter way to study geometry than the Elements, to which he replied that there was no royal road to geometry. He is therefore younger than Plato's circle, but older than Eratosthenes and Archimedes ... he was a Platonist, being in sympathy with this philosophy, whence he made the end of the whole Elements the construction of the so-called Platonic figures [regular solids].'

step, to provide proofs of increasingly sophisticated geometrical theorems. For example, Book I Proposition 5 proves that the angles at the base of an isosceles triangle (one with two equal sides) are equal. This theorem was known to generations of Victorian schoolboys as the *pons asinorum* or bridge of asses: the diagram looks like a bridge, and it was the first serious stumbling block for students who tried to learn the subject by rote instead of understanding it. Book I Proposition 32 proves that the angles of a triangle add up to 180°. Book I Proposition 47 is Pythagoras's Theorem.

Euclid deduced each theorem from previous theorems and various axioms. He built a logical tower, which climbed higher and higher towards the sky, with the axioms as its foundations and logical deduction as the mortar that bound the bricks together.

Today we are less satisfied with Euclid's logic, because it has many gaps. Euclid takes a lot of things for granted; his list of axioms is nowhere near complete. For example, it may seem obvious that if a line passes through a point inside a circle then it must cut the circle somewhere – at least if it is extended far enough. It certainly looks obvious if you draw a picture, but there are examples showing that it does not follow from Euclid's axioms. Euclid did pretty well, but he assumed that apparently obvious features of diagrams needed neither proof nor an axiomatic basis.

This omission is more serious than it might seem. There are some famous examples of fallacious reasoning arising from subtle errors in pictures. One of them 'proves' that every triangle has two equal sides.

The golden mean

Book V of the *Elements* heads off in a very different, and rather obscure, direction from Books I–IV. It doesn't look like conventional geometry. In fact, at first sight it mostly reads like gobbledegook. What, for instance, are we to make of Book V Proposition 1? It reads: *If certain magnitudes are equimultiples of other magnitudes, then whatever multiple one of the magnitudes is of one of the others, that multiple also will be of all.*

The language (which I have simplified a little) doesn't help, but the proof makes it clear what Euclid intended. The 19th-century English mathematician Augustus De Morgan explained the idea in simple language in his geometry textbook: 'Ten feet ten inches makes ten times as much as one foot one inch.'

What is Euclid up to here? Is it trivialities dressed up as theorems? Mystical nonsense? Not at all. This material may seem obscure, but it leads up to the most profound part of the *Elements*: Eudoxus's techniques for dealing with irrational ratios. Nowadays mathematicians prefer to work with numbers, and because these are more familiar I will often interpret the Greek ideas in that language.

Euclid could not avoid facing up to the difficulties of irrational numbers, because the climax to the *Elements* – and, many believe, its main objective – was the proof that there exist precisely five regular solids: the tetrahedron, cube (or hexahedron), octahedron, dodecahedron and icosahedron. Euclid proved two things: there are no other regular solids, and these five actually exist – they can be constructed geometrically, and their faces fit together perfectly, with no tiny errors.

Two of the regular solids, the dodecahedron and the icosahedron, involve the regular pentagon: the dodecahedron has pentagonal faces, and the five faces of the icosahedron surrounding any vertex determine a pentagon. Regular pentagons are directly connected with what Euclid called 'extreme and mean ratio'. On a line AB, construct a point C so that the ratio $AB{:}AC$ is equal to $AC{:}BC$. That is, the whole line bears the same proportion to the larger segment as the larger segment does to the smaller. If you draw a pentagon and inscribe a five-pointed star, the edges of the star are related to the edges of the pentagon by this particular ratio.

Nowadays we call this ratio the *golden* mean. It is equal to $\frac{1+\sqrt{5}}{2}$, and this number is irrational. Its numerical value is roughly 1.618. The Greeks could prove it was irrational by exploiting the geometry

The ratio of the diagonals to the sides is golden

Extreme and mean ratio (now called the golden mean). The ratio of the top line to the middle one is equal to that of the middle one to the bottom one

of the pentagon. So Euclid and his predecessors were aware that, for a proper understanding of the dodecahedron and icosahedron, they must come to grips with irrationals.

This, at least, is the conventional view of the *Elements*. David Fowler argues in his book *The Mathematics of Plato's Academy* that there is an alternative view – essentially, the other way round. Perhaps Euclid's main objective was the theory of irrationals, and the regular solids were just a neat application. The evidence can be interpreted either way, but one feature of the *Elements* fits this alternative theory more tidily. Much of the material on number theory is not needed for the classification of the regular solids – so why did Euclid include this material? However, the same material is closely related to irrational numbers, which could explain why it was included.

Archimedes

The greatest of the ancient mathematicians was Archimedes. He made important contributions to geometry, he was at the forefront of applications of mathematics to the natural world, and he was an accomplished engineer. But to mathematicians, Archimedes will always be remembered for his work on circles, spheres and cylinders, which we now associate with the number π ('pi'), which is roughly 3.14159. Of course the Greeks did not work with π directly: they viewed it geometrically as the ratio of the circumference of a circle to its diameter.

Earlier cultures had realized that the circumference of a circle is always the same multiple of its diameter, and knew that this multiple was roughly 3, maybe a bit bigger. The Babylonians used $3\frac{1}{8}$. But Archimedes went much further; his results were accompanied by rigorous proofs, in the spirit of Eudoxus. As far as the Greeks knew, the ratio of circumference of a circle to its diameter might be irrational. We now know that this is indeed the case, but the proof had to wait until 1770, when one was devised by Johann Heinrich Lambert. (The school value of $3\frac{1}{7}$ is convenient, but approximate.)

Be that as it may, since Archimedes could not prove π to be rational, he had to assume that it might not be.

Greek geometry worked best with polygons – shapes formed by straight lines. But a circle is curved, so Archimedes sneaked up on it by way of approximating polygons. To estimate π, he compared the circumference of a circle with the perimeters of two series of polygons: one series situated inside the circle, the other surrounding it. The perimeters of polygons inside the circle must be shorter than the circle, whereas those outside the circle must be longer than the circle. To make the calculations easier, Archimedes constructed his polygons by repeatedly bisecting the sides of a regular hexagon (six-sided polygon) getting regular polygons with 12 sides, 24, 48 and so on. He stopped at 96. His calculations proved that $3\frac{10}{71} < \pi < 3\frac{1}{7}$; that is, π lies somewhere between 3.1408 and 3.1429 in today's decimal notation.

Archimedes's work on the sphere is of special interest, because we know not just his rigorous proof, but how he found it – which was decidedly non-rigorous. The proof is given in his book *On the Sphere and Cylinder*. He shows that the volume of a sphere is two-thirds that of a circumscribed cylinder, and that the surface areas of

Pi to Vast Accuracy

The value of π has now been calculated to several billion digits, using more sophisticated methods. Such computations are of interest for their methods, to test computer systems, and for pure curiosity, but the result itself has little significance. Practical applications of π generally require no more than five or six digits. The current record is 1.24 trillion decimal digits, computed by Yasumasa Kanada and a team of nine other people in December 2002. The computation took 600 hours on a Hitachi SR8000 supercomputer.

Archimedes of Syracuse

287–212 BC

Archimedes was born in Syracuse, Greece, son of the astronomer Phidias. He visited Egypt, where supposedly he invented the Archimedean screw, which until very recently was still widely used to raise Nile water for irrigation. He probably visited Euclid in Alexandria; he definitely corresponded with Alexandrian mathematicians.

His mathematical skills were unsurpassed and wide-ranging. He turned them to practical use, and constructed gigantic war machines based on his 'law of the lever', able to hurl huge rocks at the enemy. His machines were used to good effect in the Roman siege of Alexandria in 212 BC. He even used the geometry of optical reflection to focus the Sun's rays on to an invading Roman fleet, burning the ships.

His surviving books (in later copies only) are *On Plane Equilibria, Quadrature of the Parabola, On the Sphere and Cylinder, On Spirals, On Conoids and Spheroids, On Floating Bodies, Measurement of a Circle* and *The Sandreckoner*, together with *The Method*, found in 1906 by Johan Heiberg.

those parts of the sphere and the cylinder that lie between any two parallel planes are equal. In modern language, Archimedes proved that the volume of a sphere is $\frac{4}{3}\pi r^3$, where r is the radius, and its surface area is $4\pi r^2$. These basic facts are still in use today.

The proof is an accomplished use of exhaustion. This method has an important limitation: you have to know what the answer is before you have much chance of proving it. For centuries, scholars had no idea how Archimedes guessed the answer. But in 1906 the Danish scholar Heiberg was studying a 13th-century parchment, with prayers written on it. He noticed faint lines from an earlier inscription, which had been erased to make room for the prayers.

He discovered that the original document was a copy of several works by Archimedes, some of them previously unknown. Such a document is called a palimpsest – a piece of parchment which has later writing superimposed on erased earlier writing. (Astonishingly, the same manuscript is now known to contain pieces of lost works by two other ancient authors.) One work by Archimedes, the *Method of Mechanical Theorems*, explains how to guess the volume of a sphere. The idea is to slice the sphere infinitely thinly, and place the slices at one end of a balance; at the other end, similar slices of a cylinder and a cone – whose volumes Archimedes already knew – are hung. The law of the lever produces the required value for the volume. The parchment sold for two million dollars in 1998 to a private buyer.

Problems for the Greeks

Greek geometry had limitations, some of which it overcame by introducing new methods and concepts. Euclid in effect restricted the permitted geometrical constructions to those that could be performed using an unmarked straight edge (ruler) and a pair of compasses (henceforth 'compass' – the word 'pair' is technically needed, for the same reason that we cut paper with a *pair* of

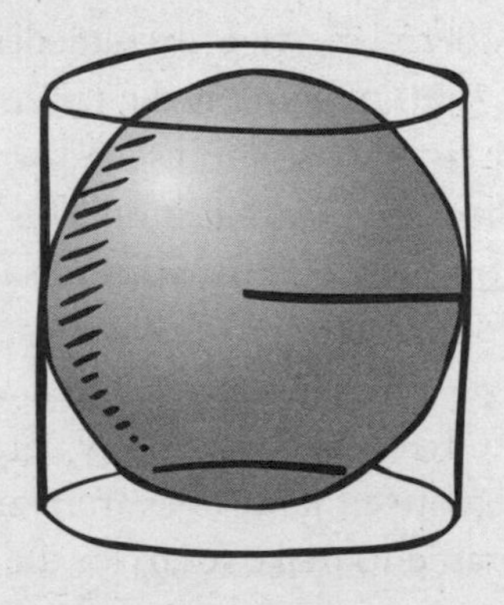

A sphere and its circumscribed cylinder

scissors, but let's not be pedantic.) It is sometimes said that he made this a requirement, but it is implicit in his constructions, not an explicit rule. With extra instruments – idealized in the same way that the curve drawn by a compass is idealized to a perfect circle – new constructions are possible.

For example, Archimedes knew that you can trisect an angle using a straight edge with two fixed marks on it. The Greeks called such procedures '*neusis* constructions'. We now know (as the Greeks must have suspected) that an exact trisection of the angle with ruler and compass is impossible, so Archimedes's contribution genuinely extends what is possible. Two other famous problems from the period are duplicating the cube (constructing a cube whose volume is twice that of a given cube) and squaring the circle (constructing a square with the same area as a given circle). These are also known to be impossible using ruler and compass.

A far-reaching extension of the allowed operations in geometry, which bore fruit in Arab work on the cubic equation around AD800 and had major applications to mechanics and astronomy, was the introduction of a new class of curves, *conic sections*. These curves, which are extraordinarily important in the history of mathematics, are obtained by slicing a double-cone with a plane. Today we shorten the name to conics. They come in three main types:

- The *ellipse*, a closed oval curve obtained when the plane meets only one half of the cone. Circles are special ellipses.
- The *hyperbola*, a curve with two infinite branches, obtained when the plane meets both halves of the cone.
- The *parabola*, a transitional curve lying between ellipses and hyperbolas, in the sense that it is parallel to some line passing through the vertex of the cone and lying on the cone. A parabola has only one branch, but extends to infinity.

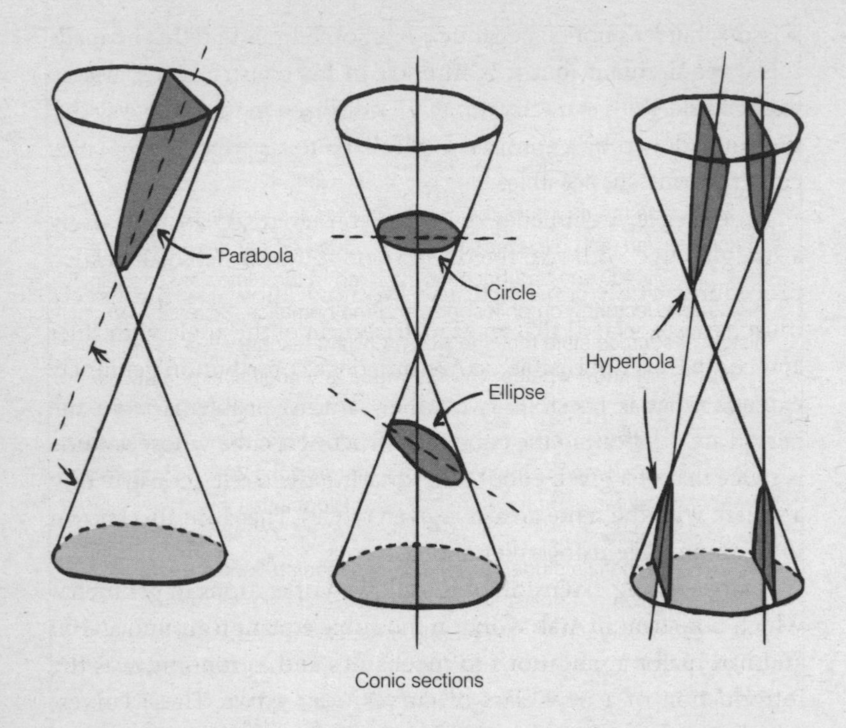

Conic sections

Conic sections were studied in detail by Apollonius of Perga, who travelled from Perga in Asia Minor to Alexandria to study under Euclid. His masterwork, the *Conic Sections* of about 230 BC, contains 487 theorems. Euclid and Archimedes had studied some properties of cones, but it would take an entire book to summarize Apollonius's theorems. One important idea deserves mention here. This is the notion of the *foci* (plural of *focus*) of an ellipse (or hyperbola). The foci are two special points associated with these two types of conic. Among their many properties, we single out just one: the distance from one focus of an ellipse, to any point, and back to the other focus is constant (equal to the long diameter of the ellipse). The foci of a hyperbola have a similar property, but now we take the difference of the two lengths.

Hypatia of Alexandria

AD370–415

Hypatia is the first woman mathematician in the historical record. She was the daughter of Theon of Alexandria, himself a mathematician, and it is probable that she learned her mathematics from him. By 400 she had become the head of the Platonist school in Alexandria, lecturing on philosophy and mathematics. Several historical sources state that she was a brilliant teacher.

We do not know whether Hypatia made any original contributions to mathematics, but she helped Theon to write a commentary on Ptolemy's *Almagest*, and may also have helped him to prepare a new edition of the *Elements*, upon which all later editions were based. She wrote commentaries on the *Arithmetica* of Diophantus and the *Conics* of Apollonius.

Among Hypatia's students were several leading figures in the growing religion of Christianity, among them Synesius of Cyrene. Some of his letters to her are recorded, and these praise her abilities. Unfortunately, many early Christians considered Hypatia's philosophy and science to be rooted in paganism, leading some to resent her influence. In 412 the new patriarch of Alexandria, Cyril, engaged in political rivalry with the Roman prefect Orestes. Hypatia was a good friend of Orestes, and her abilities as a teacher and orator were seen as a threat by the Christians. She became a focus for political unrest, and was dismembered by a mob. One source blames a fundamentalist sect, the Nitrian monks, who supported Cyril. Another blames an Alexandrian mob. A third source claims that she was part of a political rebellion, and her death was unavoidable.

Her death was brutal, hacked to pieces by a mob wielding sharp tiles (some say oyster-shells). Her mangled body was then burned. This punishment may be evidence that Hypatia was condemned for witchcraft – indeed, the first prominent witch to be killed by the early Christians – because the penalty for witchcraft prescribed by Constantius II was for their flesh to be 'torn off their bones with iron hooks'.

The Greeks knew how to trisect angles and duplicate the cube using conics. With the aid of other special curves, notably the quadratrix, they could also square the circle.

Greek mathematics contributed two crucial ideas to human development. The more obvious was a systematic understanding of geometry. Using geometry as a tool, the Greeks understood the size and shape of our planet, its relation to the Sun and Moon, even the complex motions of the remainder of the solar system. They used geometry to dig long tunnels from both ends, meeting in the middle, which cut construction time in half. They built gigantic and powerful machines, based on simple principles like the law of the lever, for purposes both peaceful and warlike. They exploited geometry in ship-building and in architecture, where buildings like the Parthenon prove to us that mathematics and beauty are not so far apart. The Parthenon's visual elegance derives from a host of clever mathematical tricks, used by the architect to overcome limitations of the human visual system and irregularities in the very ground on which the building rested.

The second Greek contribution was the systematic use of logical deduction to make sure that what was being asserted could also be justified. Logical argument emerged from their philosophy, but it found its most developed and explicit form in the geometry of Euclid and his successors. Without solid logical foundations, later mathematics could not have arisen.

Both influences remain vital today. Modern engineering – computer-based design and manufacture, for example – rests heavily on the geometric principles discovered by the Greeks. Every building is designed so that it doesn't fall down of its own accord; many are designed to resist earthquakes. Every tower block, every suspension bridge, every football stadium is a tribute to the geometers of ancient Greece.

Rational thinking, logical argument, is equally vital. Our world is too complex, potentially too dangerous, for us to base decisions on

What geometry did for them

Around 250 BC Eratosthenes of Cyrene used geometry to estimate the size of the Earth. He noticed that at midday on the summer solstice, the Sun was almost exactly overhead at Syene (present-day Aswan), because it shone straight down a vertical well. On the same day of the year, the shadow of a tall column indicated that the Sun's position at Alexandria was one fiftieth of a full circle (about 7.2°) away from the vertical. The Greeks knew that the Earth is spherical, and Alexandria was almost due north of Syene, so the geometry of a circular section of the sphere implied that the distance from Alexandria to Syene is one fiftieth of the circumference of the Earth.

Eratosthenes knew that camel trains took 50 days to get from Alexandria to Syene, and they travelled a distance of 100 stadia each day. So the distance from Alexandria to Syene is 5000 stadia, making the circumference of the Earth 250,000 stadia. Unfortunately we don't know for sure how long a stadium was, but one estimate is 157 metres, leading to a circumference of 39,250 km (24,500 miles). The modern figure is 39,840 km (24,900 miles).

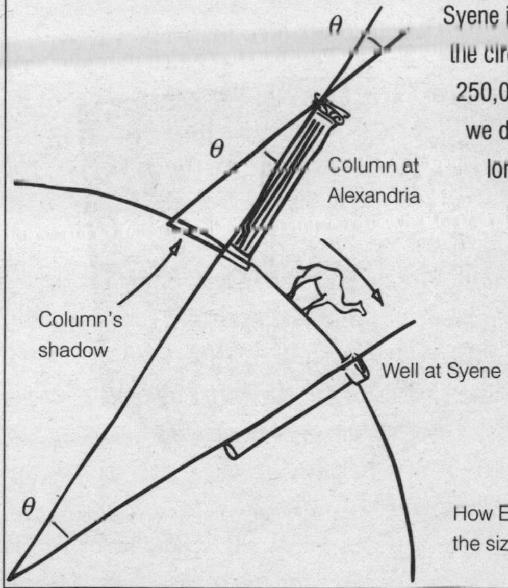

How Eratosthenes measured the size of the Earth

What geometry does for us

Archimedes's expression for the volume of a sphere is still useful today. One application, which requires knowing π to high accuracy, is the standardized unit of mass for the whole of science. For many years, for example, a metre was defined to be the length of a particular metal bar when measured at a particular temperature.

Many basic units of measurement are now defined in terms of such things as how long it takes an atom of some specific element to vibrate some huge number of times. But some are still based on physical objects, and mass is a case in point. The standard unit of mass is the kilogram. One kilogram is currently defined to be the mass of a particular sphere, made of pure silicon and kept in Paris. The sphere has been machined to extremely high accuracy. The density of silicon has also been measured very precisely. Archimedes's formula is needed to compute the volume of the sphere, which relates density to mass.

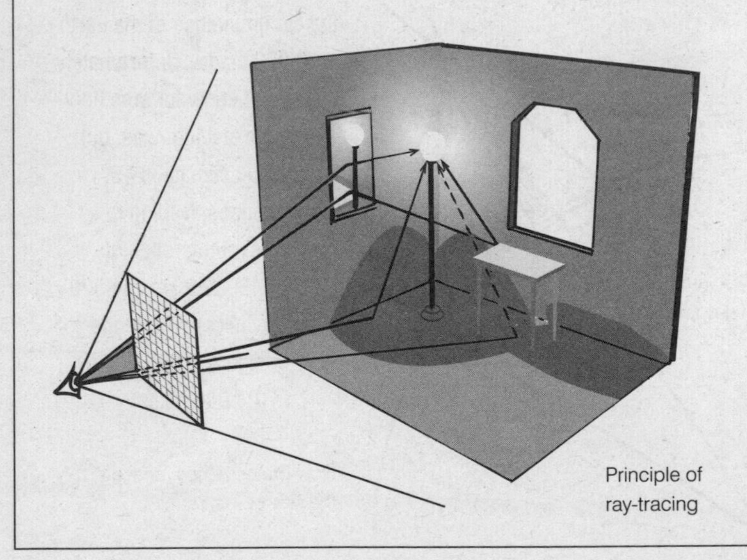

Principle of ray-tracing

> Another modern use of geometry occurs in computer graphics. Movies make extensive use of computer-generated images (CGI), and it is often necessary to generate images that include reflections – in a mirror, a wineglass, anything that catches the light. Without such reflections the image will not appear realistic. An efficient way to do this is by ray-tracing. When we look at a scene from some particular direction, our eye detects a ray of light that has bounded around the objects in the scene and happens to enter the eye from that direction. We can follow the path of this ray by working backwards. At any reflecting surface, the ray bounces so that the original ray and the reflected ray make equal angles at the surface. Translating this geometric fact into numerical calculations allows the computer to trace the ray backwards through however many bounces might be needed before it meets something opaque. (Several bounces may be necessary – if, for example, the wine glass is sitting in front of a mirror.)

what we want to believe, rather than on what is actually the case. The scientific method is deliberately constructed to overcome a deep-seated human wish to assume that what we want to be true – what we claim to 'know' – really is true. In science, emphasis is placed on trying to prove that what you deeply believe to be the case is wrong. Ideas that survive stringent attempts to disprove them are more likely to be correct.

CHAPTER 3

Notations and Numbers

Where our number symbols come from

We are so accustomed to today's number system, with its use of the ten decimal digits 0, 1, 2, 3, 4, 5, 6, 7, 8 and 9 (in Western countries), that it can come as a shock to realize that there are entirely different ways to write numbers. Even today, many cultures – Arabic, Chinese, Korean – use different symbols for the ten digits, although they all combine these symbols to form larger numbers using the same 'positional' method (hundreds, tens, units). But differences in notation can be more radical than that. There is nothing special about the number 10. It happens to be the number of human fingers and thumbs, which are ideal for counting, but if we had evolved seven fingers, or twelve, very similar systems would have worked equally well, perhaps better in some cases.

Roman numerals

Most Westerners know of at least one alternative system, Roman numerals, in which – for example – the year 2012 is written MMXII. Most of us are also aware, at least if reminded, that we employ two distinct methods for writing numbers that are not whole numbers – fractions like ¾ and decimals such as 0.75. Yet another number notation, found on calculators, is the scientific notation for very large or very small numbers – such as 5×10^9 for five billion (often seen as 5E9 on calculator displays) or 5×10^{-6} for five millionths.

These symbolic systems developed over thousands of years, and many alternatives flourished in various cultures. We have already encountered the Babylonian sexagesimal system (which would come naturally to any creature that had 60 fingers), and the simpler and more limited Egyptian number symbols, with their strange treatment of fractions. Later, base-20 numbers were used in Central America by the Mayan civilization. Only recently did humanity settle on the current methods for writing numbers, and their use became established through a mixture of tradition and convenience. Mathematics is about concepts, not symbols – but a good choice of symbol can be very helpful.

Greek numerals

We pick up the story of number symbols with the Greeks. Greek geometry was a big improvement over Babylonian geometry, but Greek arithmetic – as far as we can tell from surviving sources – was not. The Greeks took a big step backwards; they did not use positional notation. Instead, they used specific symbols for multiples of 10 or 100, so that, for instance, the symbol for 50 bore no particular relationship to that for 5 or 500.

The earliest evidence of Greek numerals is from about 1100 BC. By 600 BC the symbols had changed, and by 450 BC they had changed again, with the adoption of the Attic system, which resembles Roman numerals. The Attic system used I, II, III and IIII for the numbers 1, 2, 3 and 4. For 5 the Greek capital letter pi (Π) was employed, probably because it is the first letter of penta. Similarly, 10 was written Δ, the first letter of deka; 100 was written H, the first letter of hekaton; 1000 was written Ξ, the first letter of chilioi; and

10,000 was written M, the first letter of myrioi. Later Π was changed to Γ. So the number 2178, for example, was written as

ΞΞΗΔΔΔΔΔΔΔΓ|||

Although the Pythagoreans made numbers the basis of their philosophy, it is not known how they wrote them. Their interest in square and triangular numbers suggests that they may have represented numbers by patterns of dots. By the classical period, 600–300 BC, the Greek system had changed again, and the 27 different letters of their alphabet were used to denote numbers from 1 to 900, like this:

1	2	3	4	5	6	7	8	9
α	β	γ	δ	ε	ϛ	ζ	η	θ
10	20	30	40	50	60	70	80	90
ι	κ	λ	μ	ν	ξ	ο	π	ϙ
100	200	300	400	500	600	700	800	900
ρ	σ	τ	υ	φ	χ	ψ	ω	ϡ

These are the lower-case Greek letters, augmented by three extra letters derived from the Phoenician alphabet: ϛ (stigma), ϙ (koppa), and ϡ (sampi).

Using letters to stand for numbers might have caused ambiguity, so a horizontal line was placed over the top of the number symbols. For numbers bigger than 999, the value of a symbol could be multiplied by 1000 by placing a stroke in front of it.

The various Greek systems were reasonable as a method for recording the results of calculations, but not for performing the calculations themselves. (Imagine trying to multiply σμγ by ωλδ, for instance.) The calculations themselves were probably carried out using an abacus, perhaps represented by pebbles in the sand, especially early on.

The Greeks wrote fractions in several ways. One was to write the numerator, followed by a prime ('), and then the denominator, followed by a double prime ("). Often the denominator was written twice. So $^{21}/_{47}$ would be written as

κα' μζ" μζ",

where κα is 21 and μζ is 47. They also used Egyptian-style fractions, and there was a special symbol for $^{1}/_{2}$. Some Greek astronomers, notably Ptolemy, employed the Babylonian sexagesimal system for precision, but using Greek symbols for the component digits. It was all very different from what we use today. In fact, it was a mess.

Indian number symbols

The ten symbols currently used to denote decimal digits are often referred to as *Hindu–Arabic* numerals, because they originated in India and were taken up and developed by the Arabs.

The earliest Indian numerals were more like the Egyptian system. For example, Khasrosthi numerals, used from 400 BC to AD 100, represented the numbers from 1 to 8 as

| || ||| X |X ||X |||X XX

with a special symbol for 10. The first traces of what eventually became the modern symbolic system appeared around 300 BC in the Brahmi numerals. Buddhist inscriptions from the time include precursors of the later Hindu symbols for 1, 4 and 6. However, the Brahmi system used different symbols for multiples of ten or multiples of 100, so it was similar to the Greek number symbolism, except that it used special symbols rather than letters of the alphabet. The Brahmi system was not a positional system. By AD 100 there are records of the full Brahmi system. Inscriptions in caves and on coins show that it continued in use until the fourth century.

Between the fourth and sixth centuries, the Gupta Empire gained control of a large part of India, and the Brahmi numerals developed into Gupta numerals. From there they developed into Nagari numerals. The idea was the same, but the symbols differed.

The Indians may have developed positional notation by the first century, but the earliest datable documentary evidence for positional notation places it in 594. The evidence is a legal document which bears the date 346 in the Chedii calendar, but some scholars believe this date may be a forgery. Nevertheless, it is generally agreed that positional notation was in use in India from about 400 onwards.

There is a problem with the use of only the symbols 1–9: the notation is ambiguous. What does 25 mean, for instance? It might (in our notation) mean 25, or 205, or 2005 or 250, etc. In positional notation, where the meaning of a symbol depends on its location, it is important to specify that location without ambiguity. Today we do that by using a tenth symbol, zero (0). But it took early civilizations a long time to recognize the problem and solve it in that manner. One reason was philosophical: how can 0 be a number when a number is a quantity of things? Is nothing a quantity? Another was practical: usually it was clear from the context whether 25 meant 25 or 250 or whatever.

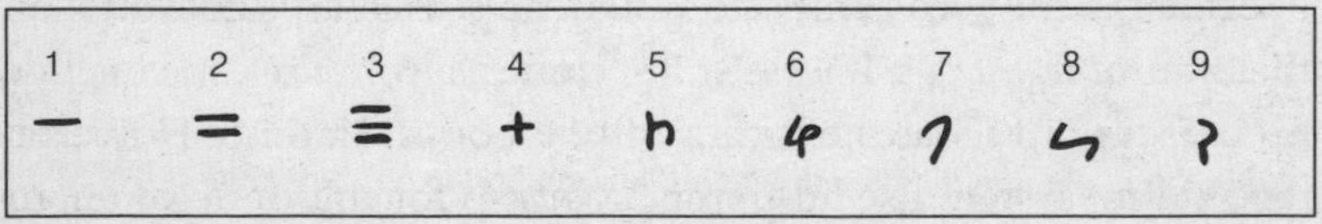

1	2	3	4	5	6	7	8	9
—	=	≡	+	[illegible]	[illegible]	[illegible]	[illegible]	[illegible]

Brahmi numerals 1–9

Some time before 400 BC – the exact date is unknown – the Babylonians introduced a special symbol to show a missing position in their number notation. This saved the scribes the effort of leaving a carefully judged space, and made it possible to work out what a number meant even if it was written sloppily. This invention was forgotten, or not transmitted to other cultures, and eventually rediscovered by the Hindus. The Bakhshali manuscript, the date of which is disputed but lies somewhere between AD 200 and 1100, uses a heavy dot •. The Jain text *Lokavibhaaga* of AD 458 uses the concept of zero, but not a symbol. A positional system that lacked the numeral zero was introduced by Aryabhata around AD 500. Later Indian mathematicians had names for zero, but did not use a symbol. The first undisputed use of zero in positional notation occurs on a stone tablet in Gwalior dated to AD 876.

Brahmagupta, Mahavira and Bhaskara

The key Indian mathematicians were Aryabhata (born AD 476), Brahmagupta (born AD 598), Mahavira (9th century) and Bhaskara (born 1114). Actually they should be described as astronomers, because mathematics was then considered to be an astronomical technique. What mathematics existed was written down as chapters in astronomy texts; it was not viewed as a subject in its own right.

Aryabhata tells us that his book *Aryabhatiya* was written when he was 23 years old. Brief though the mathematical section of his book is, it contains a wealth of material: an alphabetic system of numerals, arithmetical rules, solution methods for linear and quadratic equations, trigonometry (including the sine function and the 'versed sine' $1 - \cos\theta$). There is also an excellent approximation, 3.1416, to π.

Brahmagupta was the author of two books: *Brahma Sputa Siddhanta* and *Khanda Khadyaka*. The first is the most important; it is an astronomy text with several sections on mathematics, with arithmetic and the

What arithmetic did for them

The oldest surviving Chinese mathematics text is the *Chiu Chang*, which dates from about AD100. A typical problem is: Two and a half piculs of rice are bought for $^3/_7$ of a tael of silver. How many piculs can be bought for 9 taels? The proposed solution uses what medieval mathematicians called the 'rule of three'. In modern notation, let x be the required quantity. Then

$$\frac{x}{9} = \frac{^5/_2}{^3/_7}$$

so $x = 52\frac{1}{2}$ piculs. A picul is about 65 kilograms.

verbal equivalent of simple algebra. The second book includes a remarkable method for interpolating sine tables – that is, finding the sine of an angle from the sines of a larger angle and a smaller one.

Mahavira was a Jain, and he included a lot of Jain mathematics in his *Ganita Sara Samgraha*. This book included most of the contents of those of Aryabhata and Brahmagupta, but went a great deal further and was generally more sophisticated. It included fractions, permutations and combinations, the solution of quadratic equations, Pythagorean triangles and an attempt to find the area and perimeter of an ellipse.

Bhaskara (known as 'the teacher') wrote three important works: *Lilavati*, *Bijaganita* and *Siddhanta Siromani*. According to Fyzi, court poet of the Mogul emperor Akbar, Lilavati was the name of Bhaskara's daughter. Her father cast his daughter's horoscope, and determined the most auspicious time for her wedding. To dramatize his forecast, he put a cup with a hole in it inside a bowl of water, constructed so that it would sink when the propitious moment arrived. But Lilavati leaned over the bowl and a pearl from her clothing fell into the cup and blocked the hole. The cup did not sink, which meant that Lilavati could never get married. To cheer her up, Bhaskara wrote a mathematics textbook for her. The legend does not record what she thought of this.

Lilavati contains sophisticated ideas in arithmetic, including the method of casting out the nines, in which numbers are replaced by the sum of their digits to check calculations. It contains similar rules for divisibility by 3, 5, 7 and 11. The role of zero as a number in its own right is made clear. *Bijaganita* is about the solution of equations. *Siddhanta Siromani* deals with trigonometry: sine tables and various trigonometric relations. So great was Bhaskara's reputation that his works were still being copied around 1800.

The Hindu System

The Hindu system started to spread into the Arabic world, before it was fully developed in its country of origin. The scholar Severus Sebokht writes of its use in Syria in 662: 'I will omit all discussion of the science of the Indians ... of their subtle discoveries in astronomy ... and of their valuable methods of calculation ... I wish only to say that this computation is done by means of nine signs.'

In 776 a traveller from India appeared at the court of the Caliph and demonstrated his prowess in the 'siddhanta' method of calculation, along with trigonometry and astronomy. The basis for the computational methods seems to have been the *Brahmasphutasiddhanta* of Brahmagupta, written in 628, but whichever book it was, it was promptly translated into Arabic.

Initially the Hindu numerals were mainly used by scholars; older methods remained in widespread use among the Arabic business community and in daily life, until about 1000. But Al-Khwarizmi's *On Calculation with Hindu Numerals* of 825 made the Hindu system widely known in the Arab world. The mathematician Al-Kindi's four-volume treatise *On the Use of the Indian Numerals* (*Ketab fi Isti'mal al-'Adad al-Hindi*) of 830 increased awareness of the possibility of performing all numerical calculations using only the ten digits.

The Dark Ages?

While Arabia and India were making significant advances in mathematics and science, Europe was comparatively stagnant, although the medieval period was not quite the 'Dark Ages' of popular conception. Some advances were made, but these were slow and not particularly radical. The pace of change began to accelerate when word of the Eastern discoveries came to Europe.

Italy lies closer to the Arabian world than most parts of Europe, so it was probably inevitable that Arab advances in mathematics made their way to Europe through Italy. Venice, Genoa and Pisa were significant trading centres, and merchants sailed from these ports to North Africa and the eastern end of the Mediterranean. They exchanged wool and European wood for silks and spices.

There was metaphorical trade in ideas as well as literal trade in goods. Arabian discoveries in science and mathematics made their way along the trade routes, often by word of mouth. As trade made Europe more prosperous, barter gave way to money, and keeping accounts and paying taxes became more complex. The period's equivalent of a pocket calculator was the abacus, a device in which beads moving on wires represented numbers. However, those numbers also had to be written down on paper, for legal purposes and for general record-keeping. So the merchants needed

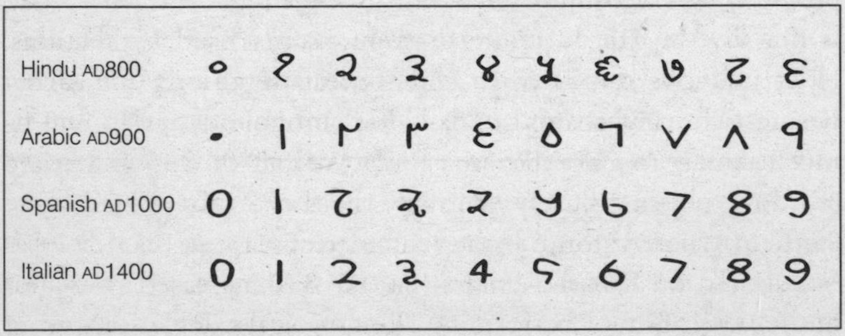

Hindu AD800

Arabic AD900

Spanish AD1000

Italian AD1400

Evolution of western number symbols

a good number notation as well as methods for doing calculations quickly and accurately.

An influential figure was Leonardo of Pisa, also known as Fibonacci, whose book *Liber Abbaci* was published in 1202. (The Italian word 'abbaco' usually means 'calculation', and need not imply the use of the abacus, a Latin term.) In this book, Leonardo introduced Hindu–Arabic number symbols to Europe.

The *Liber Abbaci* includes, and promoted, one further notational device that remains in use today: the horizontal bar in a fraction, such as $\frac{3}{4}$ for 'three-quarters'. The Hindus employed a similar notation, but without the bar; the bar seems to have been introduced by the Arabs. Fibonacci employed it widely, but his usage differed from what we do today in some respects. For instance, he would use the same bar as part of several different fractions.

Because fractions are very important in our story, it may be worth adding a few comments on the notation. In a fraction like $\frac{3}{4}$, the 4 on the bottom tells us to divide the unit into four equal parts, and the 3 on top then tells us to select three of those pieces. More formally, 4 is the *denominator* and 3 is the *numerator*. For typographical convenience, fractions are often written on a single line in the form 3/4, or sometimes in the compromise form ³⁄₄. The horizontal bar then mutates into a diagonal slash.

On the whole, however, we seldom use fractional notation in practical work. Mostly we use decimals – writing π as 3.14159, say, which is not exact, but close enough for most calculations. Historically, we have to make a bit of a leap to get to decimals, but we are following chains of ideas, not chronology, and it will be much simpler to make the leap anyway. We therefore jump forward to 1585, when William the Silent chose the Dutchman Simon Stevin as private tutor to his son Maurice of Nassau.

Building on this recognition, Stevin made quite a career for himself, becoming Inspector of Dykes, Quartermaster-General of the Army and eventually the Minister of Finance. He quickly realized

Leonardo of Pisa (Fibonacci)

1170–1250

Leonardo, Italian born, grew up in North Africa, where his father Guilielmo was working as a diplomat on behalf of merchants trading at Bugia (in modern Algeria). He accompanied his father on his numerous travels, encountered the Arabic system for writing numbers and understood its importance. In his *Liber Abbaci* of 1202 he writes: 'When my father, who had been appointed by his country as public notary in the customs at Bugia acting for the Pisan merchants going there, was in charge, he summoned me to him while I was still a child, and having an eye to usefulness and future convenience, desired me to stay there and receive instruction in the school of accounting. There, when I had been introduced to the art of the Indians' nine symbols through remarkable teaching, knowledge of the art very soon pleased me above all else.'

The book introduced the Hindu–Arabic notation to Europe, and formed a comprehensive arithmetic text, containing a wealth of material related to trade and currency conversion. Although it took several centuries for Hindu–Arabic notation to displace the traditional abacus, the advantages of a purely written system of calculation soon became apparent.

Leonardo is often known by his nickname 'Fibonacci', which means 'son of Bonaccio', but this name is not recorded before the 18th century and was probably invented then by Guillaume Libri.

the need for accurate accounting procedures, and he looked to the Italian arithmeticians of the Renaissance period, and the Hindu–Arabic notation transmitted to Europe by Leonardo of Pisa. He found fractional calculations cumbersome, and would have preferred the precision and tidiness of Babylonian sexagesimals, were it not for the use of base-60. He tried to find a system that combined the best of both, and invented a base-10 analogue of the Babylonian system: decimals.

He published his new notational system, making it clear that it had been tried, tested and found to be entirely practical by entirely practical men. In addition, he pointed out its efficacy as a business tool: 'all computations that are met in business may be performed by integers alone without the aid of fractions'.

Negative numbers

Mathematicians call the system of whole numbers the *natural numbers*. Including negative numbers as well, we obtain the integers. The *rational numbers* (or merely 'rationals') are the positive and negative fractions, the *real numbers* (or merely 'reals') are the positive and negative decimals, going on forever if necessary.

How did negative numbers come into the story?

Early in the first millennium, the Chinese employed a system of 'counting rods' instead of an abacus. They laid the rods out in patterns to represent numbers.

The top row of the picture shows *heng* rods, which represented units, hundreds, tens of thousands and so on, according to their

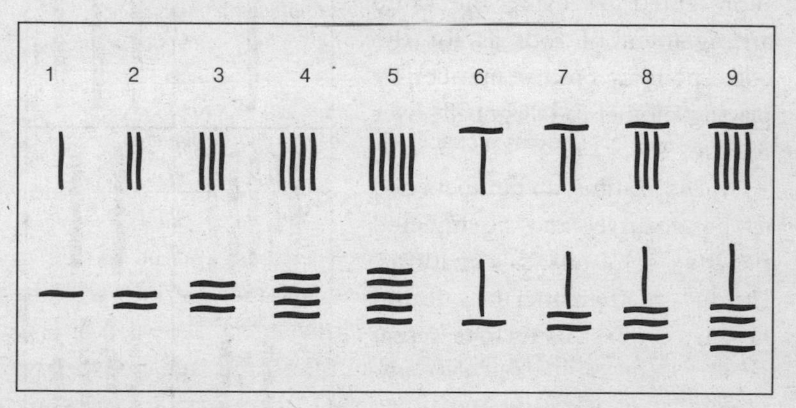

Ancient Chinese counting rods

position in a row of such symbols. The bottom row shows *tsung* rods, which represented tens, thousands and so on. So the two types alternated. Calculations were performed by manipulating the rods in systematic ways.

When solving a system of linear equations, the Chinese calculators would arrange the rods in a table. They used red rods for terms that were supposed to be added and black rods for terms that were supposed to be subtracted. So to solve equations that we would write as

$$3x - 2y = 4$$

$$x + 5y = 7$$

they would set out the two equations as two columns of a table: one with the numbers 3 (red), 2 (black), 4 (red), and the other 1 (red), 5 (red), 7 (red).

The red/black notation was not really about negative numbers, but the operation of subtraction. However, it set the stage for a concept of negative numbers, *cheng fu shu*. Now a negative number was represented by using the same arrangement of rods as for the corresponding positive number, by placing another rod diagonally over the top.

To Diophantus, all numbers had to be positive, and he rejected negative solutions to equations. Hindu mathematicians found negative numbers useful to represent

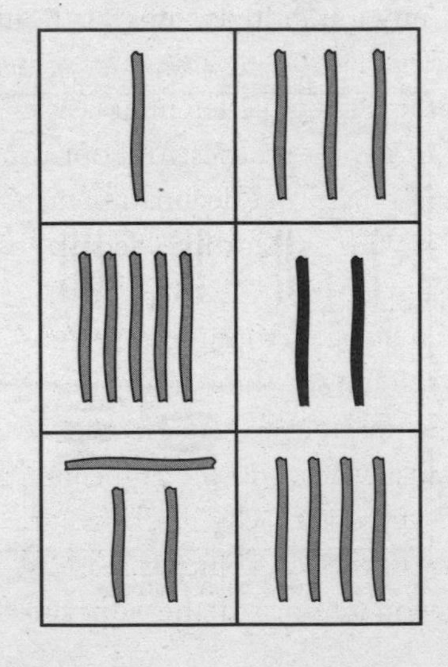

Laying out equations, Chinese style. Shaded rods are red

debts in financial calculations – owing someone a sum of money was worse, financially, than having no money, so a debt clearly should be less than zero. If you have 3 pounds and pay out 2, then you are left with $3 - 2 = 1$. By the same token, if you owe a debt of 2 pounds and acquire 3, your net worth is $-2 + 3 = 1$. Bhaskara remarks that a particular problem had two solutions, 50 and −5, but he remained nervous about the second solution, saying that it was 'not to be taken; people do not approve of negative solutions'.

Despite these misgivings, negative numbers gradually became accepted. Their interpretation, in a real calculation, needed care. Sometimes they made no sense, sometimes they might be debts, sometimes they might mean a downwards motion instead of an upwards one. But interpretation aside, their arithmetic worked perfectly well, and they were so useful as a computational aid that it would have been silly not to use them.

Arithmetic lives on

Our number system is so familiar that we tend to assume that it is the only possible one, or at least the only sensible one. In fact, it evolved, laboriously and with lots of dead ends, over thousands of years. There are many alternatives; some were used by earlier cultures, like the Mayans. Different notations for the numerals 0–9 are in use today in some countries. And our computers represent numbers internally in binary, not decimal: their programmers ensure that the numbers are turned back into decimal before they appear on the screen or in a print-out.

Since computers are now ubiquitous, is there any point in teaching arithmetic any more? Yes, for several reasons. Someone has to be able to design and build calculators and computers, and make them do the right job; this requires *understanding* arithmetic – how and why it works, not just how to do it. And if your only arithmetical ability is reading what's on a calculator, you probably won't notice if the supermarket gets your bill wrong. Without internalizing the basic operations of arithmetic, the whole of mathematics will be inaccessible to you. You might not worry about that, but modern civilization would quickly break down if we stopped teaching arithmetic, because you can't spot the future engineers and scientists at the age of five. Or even the future bank managers and accountants.

Of course, once you have a basic grasp of arithmetic by hand, using a calculator is a good way to save time and effort. But, just as you won't learn to walk by always using a crutch, you won't learn to think sensibly about numbers by relying solely on a calculator.

Mayan Numerals

A remarkable number system, which used base-20 notation instead of base-10, was developed by the Mayans, who lived in South America around 1000. In the base-20 system, the symbols equivalent to our 347 would mean

$3 \times 400 + 4 \times 20 + 7 \times 1$

(Since $20 \times 20 = 400$)

which is 1287 in our notation. The actual symbols are shown here.

1 2 3 4 5

6 7 8 9 10

11 12 13 14 15

16 17 18 19 20

40 60 80 100 120

Early civilizations that use base-10 probably did so because humans have ten fingers (including thumbs). It has been suggested that the Mayans counted on their toes as well, which is why they used base-20.

What arithmetic does for us

We use arithmetic throughout our daily lives, in commerce, and in science. Until the development of electronic calculators and computers, we either did the calculations by hand, with pen and paper, or we used aids such as the abacus or a ready reckoner (a printed book of tables of multiples of amounts of money). Today most arithmetic goes on electronically behind the scenes – supermarket checkout tills now tell the operator how much change to give back, for instance, and banks total up what is in your account automatically, rather than getting their accountant to do it. The quantity of arithmetic 'consumed' by a typical person during the course of a single day is substantial.

Computer arithmetic is not actually carried out in decimal format. Computers use base-2, or binary, rather than base-10. In place of units, tens, hundreds, thousands and so on, computers use 1, 2, 4, 8, 16, 32, 64, 128, 256, and so on – the powers of two, each twice its predecessor. (This is why the memory card for your digital camera comes in funny sizes like 256 megabytes.) In a computer, the number 100 would be broken down as 64+32+4 and stored in the form 1100100.

CHAPTER 4

Lure of the Unknown

X marks the spot

The use of symbols in mathematics goes well beyond their appearance in notation for numbers, as a casual glance at any mathematics text will make clear. The first important step towards symbolic reasoning – as opposed to mere symbolic representation – occurred in the context of problem-solving. Numerous ancient texts, right back to the Old Babylonian period, present their readers with information about some unknown quantity, and then ask for its value. A standard formula (in the literary sense) in Babylonian tablets goes 'I found a stone but did not weigh it'. After some additional information – 'when I had added a second stone of half the weight, the total weight was 15 gin' – the student is required to calculate the weight of the original stone.

Algebra

Problems of this kind eventually gave rise to what we now call algebra, in which numbers are represented by letters. The unknown quantity is traditionally denoted by the letter x, the conditions that apply to x are stated as various mathematical formulas, and the student is taught standard methods for extracting the value of x from those formulas. For instance, the Babylonian problem above would be written as $x + {}^1/_2\, x = 15$, and we would learn how to deduce that $x = 10$.

At school level, algebra is a branch of mathematics in which unknown numbers are represented by letters, the operations of arithmetic are represented by symbols and the main task is to deduce the values of unknown quantities from equations. A typical problem in school algebra is to find an unknown number x, given the equation $x^2 + 2x = 120$. This quadratic equation has one positive solution, $x = 10$. Here $x^2 + 2x = 10^2 + 2 \times 10 = 100 + 20 = 120$. It also has one negative solution, $x = -12$. Now $x^2 + 2x = (-12)^2 + 2\times(-12) = 144 - 24 = 120$. The ancients would have accepted the positive solution, but not the negative one. Today we admit both, because in many problems negative numbers have a sensible meaning and correspond to physically feasible answers, and because the mathematics actually becomes simpler if negative numbers are permitted.

In advanced mathematics, the use of letters to represent numbers is only one tiny aspect of the subject, the context in which it got started. Algebra is about the properties of symbolic expressions in their own right; it is about structure and form, not just number. This more general view of algebra developed when mathematicians started asking general questions about school-level algebra. Instead of trying to solve specific equations, they looked at the deeper structure of the solution process itself.

How did algebra arise? What came first were the problems and methods for solving them. Only later was the symbolic notation – which we now consider to be the essence of the topic – invented. There were many notational systems, but eventually one eliminated all of its competitors. The name 'algebra' appeared in the middle of this process, and it is of Arabic origin. (The initial 'al', Arabic for 'the', indicates its origin.)

Equations

What we now call the solution of equations, in which an unknown quantity must be found from suitable information, is almost as old as arithmetic. There is indirect evidence that the Babylonians were solving quite complicated equations as early as 2000 BC, and direct evidence for solutions of simpler problems, in the form of cuneiform tablets, dating from around 1700 BC.

The surviving portion of Tablet YBC 4652, from the Old Babylonian period (1800–1600 BC), contains eleven problems for solution; the text on the tablet indicates that originally there were 22 of them. A typical question is:

'I found a stone, but did not weigh it. After I weighed out six times its weight, added 2 *gin* and added one third of one seventh [of this new weight] multiplied by 24, I weighed it. The result was 1 *ma-na*. What was the original weight of the stone?'

A weight of 1 *ma-na* is 60 *gin*.

In modern notation, we would let x be the required weight in *gin*. Then the question tells us that

$$(6x + 2) + \frac{1}{3} \times \frac{1}{7} \times 24(6x + 2) = 60$$

and standard algebraic methods lead to the answer $x = 4^1/_3$ *gin*. The tablet states this answer but gives no clear indication of how it is obtained. We can be confident that it would not have been found using symbolic methods like the ones we now use, because later tablets prescribe solution methods in terms of typical examples – 'halve this number, add the product of these two, take the square root ...' and so on.

This problem, along with the others on YBC 4652, is what we now call a *linear* equation, which indicates that the unknown x enters only to the first power. All such equations can be rewritten in the form

$$ax + b = 0$$

with solution $x = -b/a$. But in ancient times, with no concept of negative numbers and no symbolic manipulation, finding a solution was not so straightforward. Even today, many students would struggle with the problem from YBC 4652.

More interesting are *quadratic* equations, in which the unknown can also appear raised to the second power – squared. The modern formulation takes the form

$$ax^2 + bx + c = 0$$

and there is a standard formula to find the value of x. The Babylonian approach is exemplified by a problem on Tablet BM 13901:

'I have added up seven times the side of my square and eleven times the area, [getting] 6;15.' (Here 6;15 is the simplified form of Babylonian sexagesimal notation, and means 6 plus 15/60, or $6\frac{1}{4}$ in modern notation.)

The stated solution runs:

'You write down 7 and 11. You multiply 6;15 by 11, [getting] 1,8;45. You break off half of 7, [getting] 3;30 and 3;30. You multiply, [getting] 12;15. You add [this] to 1,8;45 [getting] result 1,21. This is the square of 9. You subtract 3;30, which you multiplied, from 9. Result 5;30. The reciprocal of 11 cannot be found. By what must I multiply 11 to obtain 5;30? [The answer is] 0;30, the side of the square is 0;30.'

Notice that the tablet tells its reader what to do, but not why. It is a recipe. Someone must have understood why it worked, in order to write it down in the first place, but once discovered it could then be used by anyone who was appropriately trained. We don't know whether Babylonian schools merely taught the recipe, or explained why it worked.

The recipe as stated looks very obscure, but it is easier to interpret the recipe than we might expect. The complicated numbers actually help: they make it clearer which rules are being used. To find them, we just have to be systematic. In modern notation, write

$$a = 11, b = 7, c = 6;15 = 6\frac{1}{4}$$

Then the equation takes the form

$$ax^2 + bx = c$$

with those particular values for a, b, c. We have to deduce x. The Babylonian solution tells us to:

(1) Multiply c by a, which gives ac.
(2) Divide b by 2, which is $b/2$.
(3) Square $b/2$ to get $b^2/4$.
(4) Add this to ac, which is $ac + b^2/4$.
(5) Take its square root $\sqrt{ac+b^2/4}$.
(6) Subtract $b/2$, which makes $\sqrt{ac+b^2/4} - b/2$.
(7) Divide this by a, and the answer is $x = \dfrac{\sqrt{ac+b^2/4} - b/2}{a}$.

This is equivalent to the formula

$$x = \frac{-b + \sqrt{b^2 - 4ac}}{2a}$$

that is taught today because we put the term c on the left hand side, where it becomes $-c$.

It is quite clear that the Babylonians knew that their procedure was a general one. The quoted example is too complex for the solution to be a special one, designed to suit that problem alone.

How did the Babylonians think of their method, and how did they think about it? There had to be some relatively simple idea lying behind such a complicated process. It seems plausible, though there is no direct evidence, that they had a geometric idea, completing the square. An algebraic version of this is taught today, too. We can represent the question, which for clarity we choose to write in the form $x^2 + ax = b$, as a picture:

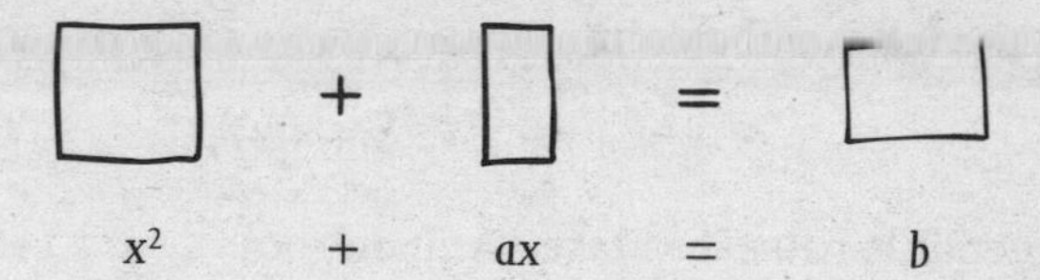

Here the square and the first rectangle have height x; their widths are, respectively, x and a. The smaller rectangle has area b. The Babylonian recipe effectively splits the first rectangle into two pieces,

$$x^2 + 2(a/2 \times x) = b$$

We can then rearrange the two new pieces and stick them on the edge of the square:

$$x^2 + 2(a/2 \times x) = b$$

The left-hand diagram now cries out to be completed to a larger square, by adding the shaded square:

To keep the equation valid, the same extra shaded square is added to the other diagram too. But now we recognize the left-hand diagram as the square of side $(x + a/2)$, and the geometric picture is equivalent to the algebraic statement

$$x^2 + 2(a/2 \times x) + (a/2)^2 = b + (a/2)^2$$

Since the left-hand side is a square, we can rewrite this as

$$(x + a/2)^2 = b + (a/2)^2$$

and then it is natural to take a square root

$$x + a/2 = \sqrt{b+(a/2)^2}$$

and finally rearrange to deduce that

$$x = \sqrt{b+(a/2)^2} - a/2$$

which is exactly how the Babylonian recipe proceeds.

There is no evidence on any tablet to support the view that this geometric picture led the Babylonians to their recipe. However, this suggestion is plausible, and is supported indirectly by various diagrams that do appear on clay tablets.

Al-jabr

The word algebra comes from the Arabic al-jabr, a term employed by Muhammad ibn Musa al-Khwarizmi, who flourished around 820. His work *The Compendious Book on Calculation by al-jabr w'al-muqabala* explained general methods for solving equations by manipulating unknown quantities.

Al-Khwarizmi used words, not symbols, but his methods are recognizably similar to those taught today. *Al-jabr* means 'adding equal amounts to both sides of an equation', which is what we do when we start from

$$x - 3 = 5$$

and deduce that

$$x = 8$$

In effect, we make this deduction by adding 3 to both sides. *Al-muqabala* has two meanings. There is a special meaning: 'subtracting equal amounts from both sides of an equation', which we do to pass from

$$x + 3 = 5$$

to the answer

$$x = 2$$

But it also has a general meaning: 'comparison'.

Al-Khwarizmi gives general rules for solving six kinds of equation, which between them can be used to solve all linear and quadratic equations. In his work, then, we find the ideas of elementary algebra, but not the use of symbols.

Cubic equations

The Babylonians could solve quadratic equations, and their method was essentially the same one that is taught today. Algebraically, it involves nothing more complicated than a square root, beyond the standard operations of arithmetic (add, subtract, multiply, divide). The obvious next step is cubic equations, involving the cube of the unknown. We write such equations as

$$ax^3 + bx^2 + cx + d = 0$$

where x is the unknown and the coefficients a, b, c, d are known numbers. But until the development of negative numbers, mathematicians classified cubic equations into many distinct types – so that, for example, $x^3 + 3x = 7$ and $x^3 - 3x = 7$ were considered to be completely different, and required different methods for their solution.

The Greeks discovered how to use conic sections to solve some cubic equations. Modern algebra shows that if a conic intersects another conic, the points of intersection are determined by an equation of third or fourth degree (depending on the conics). The Greeks did not know this as a general fact, but they exploited its consequences in specific instances, using the conics as a new kind of geometrical instrument.

This line of attack was completed and codified by the Persian Omar Khayyam, best known for his poem the *Rubaiyat*. Around 1075 he classified cubic equations into 14 kinds, and showed how to solve each kind using conics, in his work *On the Proofs of the Problems of Algebra and Muqabala*. The treatise was a geometric *tour de force*, and it polished off the geometric problem almost completely. A modern mathematician would raise a few quibbles – some of Omar's cases are not completely solved because he assumes that certain geometrically constructed points exist when sometimes they do not. That is, he assumes his conics meet when they may fail to do so. But these are minor blemishes.

Geometric solutions of cubic equations were all very well, but could there exist algebraic solutions, involving such things as cube roots, but nothing more complicated? The mathematicians of Renaissance Italy made one of the biggest breakthroughs in algebra when they discovered that the answer is 'yes'.

In those days, mathematicians made their reputation by taking part in public contests. Each contestant would set his opponent problems, and whoever solved the most was adjudged the winner. Members of the audience could place bets on who would win. The contestants often wagered large sums of money – in one recorded instance, the loser had to buy the winner (and his friends) thirty banquets. Additionally, the winner's ability to attract paying students, mostly from the nobility, was likely to be enhanced. So, public mathematical combat was serious stuff.

In 1535 there was just such a contest, between Antonio Fior and Niccolo Fontana, nicknamed Tartaglia, 'the stammerer'. Tartaglia wiped the floor with Fior, and word of his success spread, coming to the ears of Girolamo Cardano. And Cardano's ears pricked up. He was in the middle of writing a comprehensive algebra text, and the questions that Fior and Tartaglia had posed each other were – cubic equations. At that time, cubic equations were classified into three distinct types, again because negative numbers were not recognized. Fior knew how to solve just one type. Initially, Tartaglia knew how to solve just one different type. In modern symbols, his solution of a cubic equation of the type $x^3 + ax = b$ is

$$x = \sqrt[3]{\frac{b}{2} + \sqrt{\frac{a^3}{27} + \frac{b^2}{4}}} + \sqrt[3]{\frac{b}{2} - \sqrt{\frac{a^3}{27} + \frac{b^2}{4}}}$$

In a burst of inspired desperation, a week or so before the contest, Tartaglia figured out how to solve the other types too. He then set Fior only the types that he knew Fior could not solve.

Cardano, hearing of the contest, realized that the two combatants had devised methods for solving cubic equations. Wanting to add them to his book, he buttonholed Tartaglia and asked him to reveal his methods. Tartaglia was naturally reluctant, because his livelihood depended on them, but eventually he was persuaded to divulge the secret. According to Tartaglia, Cardano promised never to make the method public. So Tartaglia was understandably peeved when his method appeared in Cardano's *Ars Magna – the Great Art of Algebra*. He complained bitterly and accused Cardano of plagiarism.

Now, Cardano was far from lilywhite. He was an inveterate gambler, who had made and lost considerable sums of money at cards, dice and even chess. He lost the entire family fortune in this manner, and was reduced to penury. He was also a genius, a competent doctor, a brilliant mathematician and an accomplished self-publicist – though his positive attributes were mitigated by frankness that often became offensively blunt and insulting. So Tartaglia can be forgiven for assuming that Cardano had lied to him and stolen his discovery. That Cardano had given full credit to Tartaglia in his book only made things worse; Tartaglia knew that it was the book's author who would be remembered, not some obscure figure given a sentence or so of mention.

However, Cardano had an excuse, quite a good one. And he also had a strong reason to bend his promise to Tartaglia. The reason was that Cardano's student Lodovico Ferrari had found a method for solving quartic equations, those involving the fourth power of the unknown. This was completely new, and of huge importance. So of

Fibonacci Sequence

The third section of the *Liber Abbaci* contains a problem that seems to have originated with Leonardo: 'A certain man put a pair of rabbits in a place surrounded on all sides by a wall. How many pairs of rabbits can be produced from that pair in a year if in every month, each pair begets a new pair, which from the second month onwards becomes productive?'

This rather quirky problem leads to a curious, and famous, sequence of numbers:

1 2 3 5 8 13 21 34 55

and so on. Each number is the sum of the two preceding numbers. This is known as the *Fibonacci Sequence*, and it turns up repeatedly in mathematics and in the natural world. In particular, many flowers have a Fibonacci number of petals. This is not coincidence, but a consequence of the growth pattern of the plant and the geometry of the 'primordia' – tiny clumps of cells at the tip of the growing shoot that give rise to important structures, including petals.

Although Fibonacci's growth rule for rabbit populations is unrealistic, more general rules of a similar kind (called *Leslie models*) are used today for certain problems in population dynamics, the study of how animal populations change in size as the animals breed and die.

course Cardano wanted quartic equations in his book, too. Since it was his student who had made the discovery, this would have been legitimate. But Ferrari's method reduced the solution of any quartic to that of an associated cubic, so it relied on Tartaglia's solution of cubic equations. Cardano could not publish Ferrari's work without also publishing Tartaglia's.

Then news reached him that offered a way out. Fior, who had lost to Tartaglia in public combat, was a student of Scipio del Ferro. Cardano heard that del Ferro had solved all three types of cubic, not

What algebra did for them

Several chapters of the *Liber Abbaci* contain algebraic problems relevant to the needs of merchants. One, not terribly practical, goes like this: 'A man buys 30 birds – partridges, doves and sparrows. A partridge costs 3 silver coins, a dove 2, and a sparrow ½. He pays 30 silver coins. How many birds of each type does he buy?'

In modern notation, if we let x be the number of partridges, y the number of doves, and z the number of sparrows, then we must solve two equations

$$x + y + z = 30$$
$$3x + 2y + \tfrac{1}{2}z = 30$$

In real or rational numbers, these equations would have infinitely many solutions, but there is an extra condition implied by the question: the numbers x, y, z are integers. It turns out that only one solution exists: 3 partridges, 5 doves and 22 sparrows.

Leonardo also mentions a series of problems about buying a horse. One man says to another, 'If you give me one-third of your money, I can buy the horse'. The other says, 'if you give me one-quarter of your money, I can buy the horse'. What is the price of the horse? This time there are many solutions; the smallest one in whole numbers sets the price of the horse at 11 silver coins.

just the one that he had passed on to Fior. And a certain Annibale del Nave was rumoured to possess del Ferro's unpublished papers. So Cardano and Ferrari went to Bologna in 1543 to consult del Nave, viewed the papers – and there, as plain as the nose on your face, were solutions of all three types of cubic. So Cardano could honestly say that he was not publishing Tartaglia's method, but del Ferro's.

Tartaglia didn't see things that way. But he had no real answer to Cardano's point that the solution was not Tartaglia's discovery at all, but del Ferro's. Tartaglia published a long, bitter diatribe about the

Girolamo Cardano

(aka Hieronymus Cardanus, Jerome Cardan)
1501–1576

Girolamo Cardano was the illegitimate son of the Milanese lawyer Fazio Cardano and a young widow named Chiara Micheria who was trying to bring up three children. The children died of the plague in Milan while Chiara was giving birth to Girolamo in nearby Pavia. Fazio was an able mathematician, and he passed on his passion for the subject to Girolamo. Against his father's wishes, Girolamo studied medicine at Pavia University; Fazio had wanted him to study law.

While still a student, Cardano was elected rector of the University of Padua, to which he had moved, by a single vote. Having spent a small legacy from his recently deceased father, Cardano turned to gambling to augment his finances: cards, dice and chess. He always carried a knife and once slashed the face of an opponent whom he believed he had caught cheating.

In 1525 Cardano gained his medical degree, but his application to join the College of Physicians in Milan was rejected, probably because he had a reputation for being difficult. He practised medicine in the village of Sacca, and married Lucia Bandarini, a militia captain's daughter. The practice did not prosper, and in 1533 Girolamo again turned to gambling, but now he lost heavily, and had to pawn his wife's jewellery and some of the family furniture.

Cardano struck lucky, and was offered his father's old position as lecturer in mathematics at the Piatti Foundation. He continued practising medicine on the side, and some miraculous cures enhanced his reputation as a doctor. By 1539, after several attempts, he was finally admitted to the College of Physicians. He began to publish scholarly texts on a variety of topics, including mathematics.

Cardano wrote a remarkable autobiography, *The Book of My Life*, a miscellany of chapters on numerous topics. His fame was at its peak, and he visited Edinburgh to treat the Archbishop of St Andrews, John Hamilton. Hamilton suffered from severe asthma. Under Cardano's care, his health improved dramatically, and Cardano left Scotland 2000 gold crowns the richer.

He became professor at Pavia University, and things were going swimmingly until his eldest son Giambatista secretly married Brandonia di Seroni, 'a worthless, shameless woman' in Cardano's estimation. She and her family publicly humiliated and taunted Giambatista, who poisoned her. Despite Cardano's best efforts, Giambatista was executed. In 1570 Cardano was tried for heresy, having cast the horoscope of Jesus. He was imprisoned, then released, but banned from university employment. He went to Rome, where the Pope unexpectedly gave him a pension and he was admitted to the College of Physicians.

He forecast the date of his own death, and allegedly made sure he was right by committing suicide. Despite many tribulations, he remained an optimist to the end.

affair, and was challenged to a public debate by Ferrari, defending his master. Ferrari won hands down, and Tartaglia never really recovered from the setback.

Algebraic symbolism

The mathematicians of Renaissance Italy had developed many algebraic methods, but their notation was still rudimentary. It took hundreds of years for today's algebraic symbolism to develop.

One of the first to use symbols in place of unknown numbers was Diophantus of Alexandria. His *Arithmetica*, written around 250, originally consisted of 13 books, of which six have survived as later copies. Its focus is the solution of algebraic equations, either in whole numbers or in rational numbers – fractions p/q where p and q are whole numbers. Diophantus's notation differs considerably from what we use today. Although the *Arithmetica* is the only surviving document on this topic, there is fragmentary evidence that Diophantus was part of a wider tradition, and not just an isolated figure. Diophantus's notation is not very well suited to calculations, but it does summarize them in a compact form.

The Arabic mathematicians of the Medieval period developed sophisticated methods for solving equations, but expressed them in words, not symbols.

Diophantus's Notation and Ours

Meaning	Modern symbol	Diophantus's symbol
The unknown	x	γ
Its square	x^2	Δγ
Its cube	x^3	Kγ
Its fourth power	x^4	ΔγΔ
Its fifth power	x^5	ΔKγ
Its sixth power	x^6	KγK
Addition	+	Juxtapose terms (use AB for A+B)
Subtraction	–	⋔
Equality	=	ι^{σ}

The move to symbolic notation gained momentum in the Renaissance period. The first of the great algebraists to start using symbols was François Vieta, who stated many of his results in symbolic form, but his notation differed considerably from the modern one. He did, however, use letters of the alphabet to represent known quantities, as well as unknown ones. To distinguish these, he adopted the convention that consonants $B, C, D, F, G \ldots$ represented known quantities, whereas vowels $A, E, I, \ldots$ represented unknowns.

In the 15th century, a few rudimentary symbols made their appearance, notably the letters *p* and *m* for addition and subtraction: *plus* and *minus*. These were abbreviations rather than true symbols. The symbols $+$ and $-$ also appeared around this time. They arose in commerce, where they were used by German merchants to distinguish overweight and underweight items. Mathematicians quickly began to employ them too, the first written examples appearing in 1481. William Oughtred introduced the symbol $\times$ for multiplication, and was roundly (and rightly) criticized by Leibniz on the grounds that this was too easily confused with the letter x.

In 1557, in his *The Whetstone of Witte*, the English mathematician Robert Recorde invented the symbol $=$ for equality, in use ever since. He wrote that he could think of no two things that were more alike than two parallel lines of the same length. However, he used much longer lines than we do today, more like $=\!=\!=\!=\!=\!=$. Vieta initially wrote the word 'aequalis' for equality, but later replaced it by the symbol $\sim$. René Descartes used a different symbol, $\propto$.

The current symbols $>$ and $<$ for 'greater than' and 'less than' are due to Thomas Harriot. Round brackets ( ) show up in 1544, and square [ ] and curly { } brackets were used by Vieta around 1593. Descartes used the square root symbol $\surd$, which is an elaboration on the letter r for radix, or root; but he wrote $\surd c$ for the cube root.

To see how different Renaissance algebraic notation was from ours, here is a short extract from Cardano's *Ars Magna*:

5p: ℞ m:15

5m: ℞ m:15

25m:m:15 *qd. est* 40

In modern notation this would read:

$$(5+\sqrt{-15})(5-\sqrt{-15}) = 25-(-15) = 40$$

So here we see p: and m: for plus and minus, ℞ for 'square root', and '*qd. est*' abbreviating the Latin phrase 'which is'. He wrote

qdratu aeqtur 4 *rebus* p:32

where we would write

$$x^2 = 4x + 32$$

and therefore used separate abbreviations '*rebus*' and '*qdratu*' for the unknown (thing) and its square. Elsewhere he used R for the unknown, Z for its square and C for its cube.

An influential but little-known figure was the Frenchman Nicolas Chuquet, whose book *Triparty en la Science de Nombres* of 1484 discussed three main mathematical topics: arithmetic, roots and unknowns. His notation for roots was much like Cardano's, but he started to systematize the treatment of powers of the unknown, by using superscripts for exponents. He referred to the first four powers of the unknown as *premier*, *champs*, *cubiez* and *champs de champs*. For what we would now write as $6x$, $4x^2$ and $5x^3$ he used $.6.^1$, $.4.^2$ and $.5.^3$. He also used zeroth and negative powers, writing $.2.^0$ and $.3.^{1.m.}$ where we would write $2x^0$ and $3x^{-1}$. In short: he used exponential notation (superscripts) for powers of the unknown, but had no explicit symbol for the unknown itself.

That omission was supplied by Descartes. His notation was very similar to what we use nowadays, with one exception. Where we would write

$$5 + 4x + 6x^2 + 11x^3 + 3x^4$$

say, Descartes wrote

$$5 + 4x + 6xx + 11x^3 + 3x^4$$

That is, he used xx for the square. Occasionally, though he used x^2. Newton wrote powers of the unknown exactly as we do now, including fractional and negative exponents, such as $x^{3/2}$ for the square root of x^3. It was Gauss who finally abolished xx in favour of x^2; once the Grand Master had done this, everyone else followed suit.

The logic of species

Algebra began as a way to systematize problems in arithmetic, but by the time of Vieta it had acquired a life of its own. Before Vieta, algebraic symbolism and manipulation were viewed as ways to state and carry out arithmetical procedures, but numbers were still the main point. Vieta made a crucial distinction between what he called the logic of species and the logic of numbers. In his view, an algebraic expression represented an entire class (species) of arithmetical expressions. It was a different concept. In his 1591 *In Artem Analyticam Isagoge* (Introduction to the Analytic Art) he explained that algebra is a method for operating on general forms, whereas arithmetic is a method for operating on specific numbers.

This may sound like logical hair-splitting, but the difference in the point of view was significant. To Vieta, an algebraic calculation such as (in our notation)

$$(2x + 3y) - (x + y) = x + 2y$$

What algebra does for us

The leading consumers of algebra in the modern world are scientists, who represent nature's regularities in terms of algebraic equations. These equations can be solved to represent unknown quantities in terms of known ones. The technique has become so routine that no one notices they're using algebra.

Algebra was very nearly applied to archaeology in one episode of *Time Team*, when the intrepid TV archaeologists wanted to work out how deep a mediaeval well was. The first idea was to drop something down it, and time how long it took to reach the bottom. It took six seconds. The relevant algebraic formula here, neglecting the speed of sound, is

$$s = \tfrac{1}{2}gt^2$$

where s is the depth, t is the time taken to hit the bottom, and g is the acceleration due to gravity, roughly 10 metres per second^2. Taking $t = 6$, the formula tells us that the well is roughly 180 metres deep.

Because of some uncertainty about the formula – which in fact they had remembered correctly – the *Time Team* used three long tape-measures tied together.

The measured depth was in fact very close to 180 metres.

Algebra enters more obviously if we know the depth and want to calculate the time. Now we have to solve the equation for t in terms of s, leading to the answer

$$t = \sqrt{\frac{2s}{g}}$$

Knowing that $s = 180$ metres for instance, lets us predict that t is the square root of 360/10, that is, the square root of 36 – which is 6 seconds.

expresses a way to manipulate symbolic expressions. The individual terms $2x + 3y$ and so on are themselves mathematical objects. They can be added, subtracted, multiplied and divided without ever considering them as representations of specific numbers. To Vieta's predecessors, however, the same equation was simply a numerical relationship that was valid whenever specific numbers were substituted for the symbols x and y. So algebra took on a life of its own, as the mathematics of symbolic expressions. It was the first step towards freeing algebra from the shackles of arithmetical interpretation.

CHAPTER 5

Eternal Triangles

Trigonometry and logarithms

Euclidean geometry is based on triangles, mainly because every polygon can be built from triangles, and most other interesting shapes, such as circles and ellipses, can be approximated by polygons. The metric properties of triangles – those that can be measured, such as the lengths of sides, the sizes of angles or the total area – are related by a variety of formulas, many of them elegant. The practical use of these formulas, which are extremely useful in navigation and surveying, required the development of trigonometry, which basically means 'measuring triangles'.

Trigonometry

Trigonometry spawned a number of special functions – mathematical rules for calculating one quantity from another. These functions go by names like *sine*, *cosine* and *tangent*. The trigonometric functions turned out to be of vital importance for the whole of mathematics, not just for measuring triangles.

Trigonometry is one of the most widely used mathematical techniques, involved in everything from surveying to navigation to GPS satellite systems in cars. Its use in science and technology is so common that it usually goes unnoticed, as befits any universal tool. Historically, it was closely associated with logarithms, a clever method for converting multiplications (which are hard) into additions (which are easier). The main ideas appeared between about 1400 and 1600, though with a lengthy prehistory and plenty of later embellishments, and the notation is still evolving.

In this chapter we'll take a look at the basic topics: trigonometric functions, the exponential function and the logarithm. We also consider a few applications, old and new. Many of the older applications are computational techniques, which have mostly become obsolete now that computers are widespread. Hardly anyone now uses logarithms to do multiplication, for instance. No one uses tables at all, now that computers can rapidly calculate the values of functions to high precision. But when logarithms were first invented, it was the numerical tables of them that made them useful, especially in areas like astronomy, where long and complicated numerical calculations were necessary. And the compilers of the tables had to spend years – decades – of their lives doing the sums. Humanity owes a great deal to these dedicated and dogged pioneers.

The origins of trigonometry

The basic problem addressed by trigonometry is the calculation of properties of a triangle – lengths of sides, sizes of angles – from other such properties. It is much easier to describe the early history of trigonometry if we first summarize the main features of modern trigonometry, which is mostly a reworking in 18th century notation of topics that go right back to the Greeks, if not earlier. This summary provides a framework within which we can describe the ideas of the ancients, without getting tangled up in obscure and eventually obsolete concepts.

Trigonometry seems to have originated in astronomy, where it is relatively easy to measure angles, but difficult to measure the vast distances. The Greek astronomer Aristarchus, in a work of around

Trigonometry – a Primer

Trigonometry relies on a number of special functions, of which the most basic are the *sine*, *cosine* and *tangent*. These functions apply to an angle, traditionally represented by the Greek letter θ (theta). They can be defined in terms of a right triangle, whose three sides a, b, c are called the *adjacent* side, the *opposite* side and the *hypotenuse*.

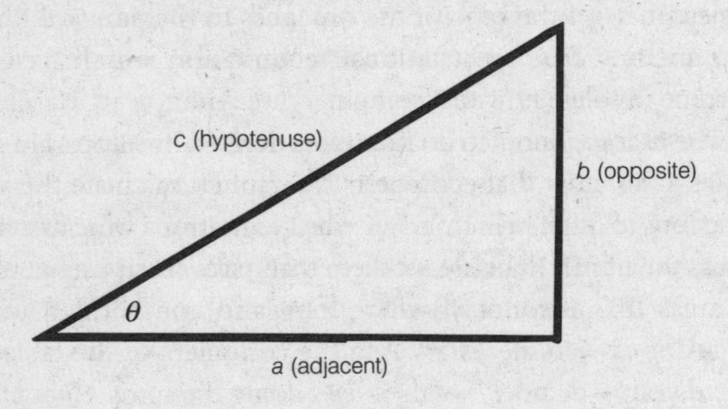

Then:

The *sine* of theta is $\sin\theta = b/c$

The *cosine* of theta is $\cos\theta = a/c$

The *tangent* of theta is $\tan\theta = b/a$

As it stands, the values of these three functions, for any given angle θ, are determined by the geometry of the triangle. (The same angle may occur in triangles of different sizes, but the geometry of similar triangles implies that the stated *ratios* are independent of size.) However, once these functions have been calculated and tabulated, they can be used to solve (calculate all the sides and angles of) the triangle from the value of θ.

The three functions are related by a number of beautiful formulas. In particular, Pythagoras's Theorem implies that

$$\sin^2\theta + \cos^2\theta = 1$$

260 BC, *On the Sizes and Distances of the Sun and Moon*, deduced that the Sun lies between 18 and 20 times as far from the Earth as the Moon does. (The correct figure is closer to 400, but Eudoxus and Phidias had argued for 10.) His reasoning was that when the Moon is half full, the angle between the directions from the observer to the Sun and the Moon is about 87° (in modern units). Using properties of triangles that amount to trigonometric estimates, he deduced (in modern notation) that sin 3° lies between $^1/_{18}$ and $^1/_{20}$, leading to his estimate of the ratio of the distances to the Sun and the Moon. The method was right, but the observation was inaccurate; the correct angle is 89.8°.

The first trigonometric tables were derived by Hipparchus around 150 BC. Instead of the modern sine function, he used a closely related quantity, which from the geometric point of view was equally natural. Imagine a circle, with two radial lines meeting at an angle θ. The points where these lines cut the circle can be joined

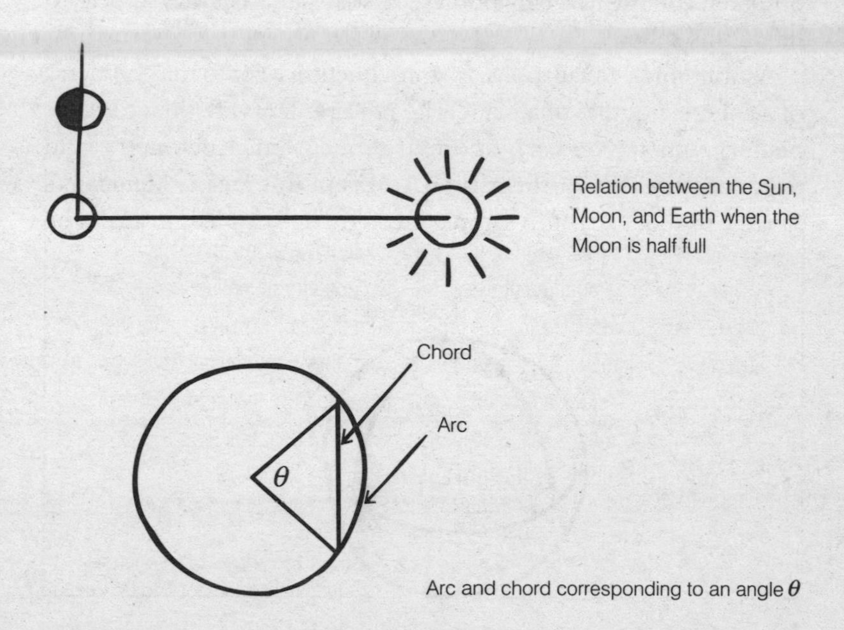

Relation between the Sun, Moon, and Earth when the Moon is half full

Arc and chord corresponding to an angle θ

by a straight line, called a *chord*. They can also be thought of as the end points of a curved part of the circle, called an *arc*.

Hipparchus drew up a table relating arc and chord length for a range of angles. If the circle has radius 1, then the arc length is equal to θ when this angle is measured in units known as *radians*. Some easy geometry shows that the chord length in modern notation is $2\sin \theta/2$. So Hipparchus's calculation is very closely related to a table of sines, even though it was not presented in that way.

Astronomy

Remarkably, early work in trigonometry was more complicated than most of what is taught in schools today, again because of the needs of astronomy (and, later, navigation). The natural space to work in was not the plane, but the sphere. Heavenly objects can be thought of as lying on an imaginary sphere, the *celestial sphere*. Effectively, the sky looks like the inside of a gigantic sphere surrounding the observer, and the heavenly bodies are so distant that they appear to lie on this sphere.

Astronomical calculations, in consequence, refer to the geometry of a sphere, not that of a plane. The requirements are therefore not plane geometry and trigonometry, but *spherical* geometry and trigonometry. One of the earliest works in this area is Menelaus's *Sphaerica* of about AD100. A sample theorem, one that has no analogue

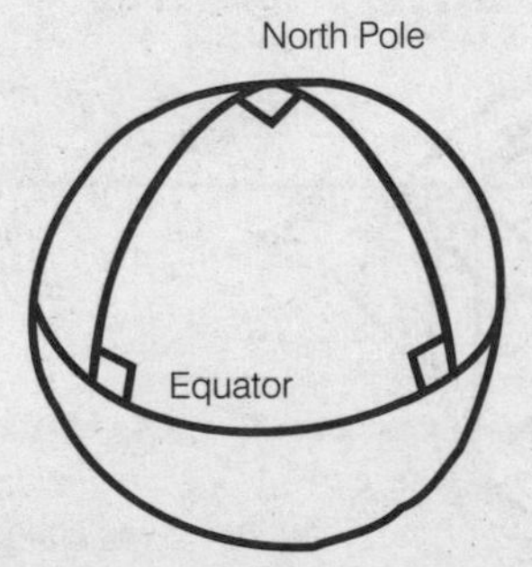

The angles of a spherical triangle do not add up to 180°

in Euclidean geometry, is this: if two triangles have the same angles as each other, then they are *congruent* – they have the same size and shape. (In the Euclidean case, they are similar – same shape but possibly different sizes.) In spherical geometry, the angles of a triangle do not add up to 180°, as they do in the plane. For example, a triangle whose vertices lie at the North Pole and at two points on the equator separated by 90° clearly has all three angles equal to a right angle, so the sum is 270°. Roughly speaking, the bigger the triangle becomes, the bigger its angle-sum becomes. In fact, this sum, minus 180°, is proportional to the triangle's total area.

These examples make it clear that spherical geometry has its own characteristic and novel features. The same goes for spherical trigonometry, but the basic quantities are still the standard trigonometric functions. Only the formulas change.

Ptolemy

By far and above the most important trigonometry text of antiquity was the *Mathematical Syntaxis* of Ptolemy of Alexandria, which dates to about AD150. It is better known as the *Almagest*, an Arabic term meaning 'the greatest'. It included trigonometric tables, again stated in terms of chords, together with the methods used to calculate them, and a catalogue of star positions on the celestial sphere. An essential feature of the computational method was Ptolemy's

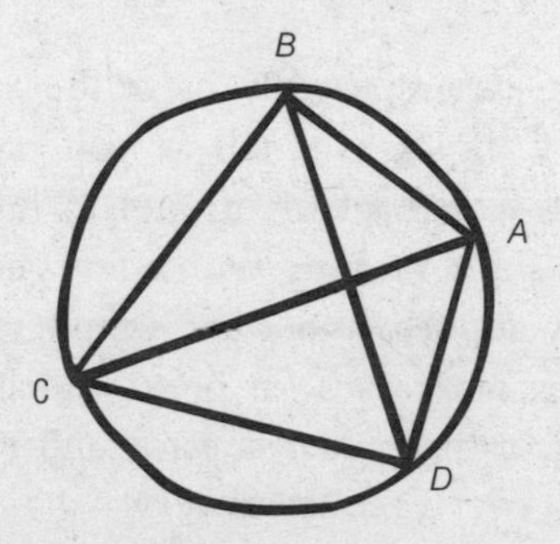

Cyclic quadrilateral and its diagonals

Theorem which states that if *ABCD* is a cyclic quadrilateral (one whose vertices lie on a circle) then

$$AB \times CD + BC \times DA = AC \times BD$$

(the sum of the products of opposite pairs of sides is equal to the product of the diagonals).

A modern interpretation of this fact is the remarkable pair of formulas

$$\sin(\theta + \varphi) = \sin\theta\cos\varphi + \cos\theta\sin\varphi$$

$$\cos(\theta + \varphi) = \cos\theta\cos\varphi - \sin\theta\sin\varphi$$

The main point about these formulas is that if you know the sines and cosines of two angles, then you can easily work the sines and cosines out for the sum of those angles. So, starting with (say) sin 1° and cos 1°, you can deduce sin 2 and cos 2° by taking $\theta = \varphi = 1°$. Then you can deduce sin 3° and cos 3° by taking $\theta = 1°$, $\varphi = 2°$, and so on. You had to know how to start, but after that, all you needed was arithmetic – rather a lot of it, but nothing more complicated.

Getting started was easier than it might seem, requiring arithmetic and square roots. Using the obvious fact that $\theta/2 + \theta/2 = \theta$, Ptolemy's Theorem implies that

$$\sin\frac{\theta}{2} = \sqrt{\frac{1 - \cos\theta}{2}}$$

Starting from cos 90° = 0, you can repeatedly halve the angle, obtaining sines and cosines of angles as small as you please. (Ptolemy used ¼°.) Then you can work back up through all integer multiples of that small angle. In short, starting with a few general trigonometric formulas, suitably applied, and a few simple values for specific angles you can work out values for pretty much any angle you want. It was an extraordinary *tour de force*, and it put astronomers in business for well over a thousand years.

A final noteworthy feature of the *Almagest* is how it handled the orbits of the planets. Anyone who watches the night sky regularly quickly discovers that the planets wander against the background of fixed stars, and that the paths they follow seem rather complicated, sometimes moving backwards, or travelling in elongated loops.

Eudoxus, responding to a request from Plato, had found a way to represent these complex motions in terms of revolving spheres mounted on other spheres. This idea was simplified by Apollonius and Hipparchus, to use *epicycles* – circles whose centres move along other circles, and so on. Ptolemy refined the system of epicycles, so that it provided a very accurate model of the planetary motions.

Early trigonometry

Early trigonometric concepts appear in the writings of Hindu mathematicians and astronomers: Varahamihira's *Pancha Siddhanta* of 500, Brahmagupta's *Brahma Sputa Siddhanta* of 628 and the more detailed *Siddhanta Siromani* of Bhaskaracharya in 1150.

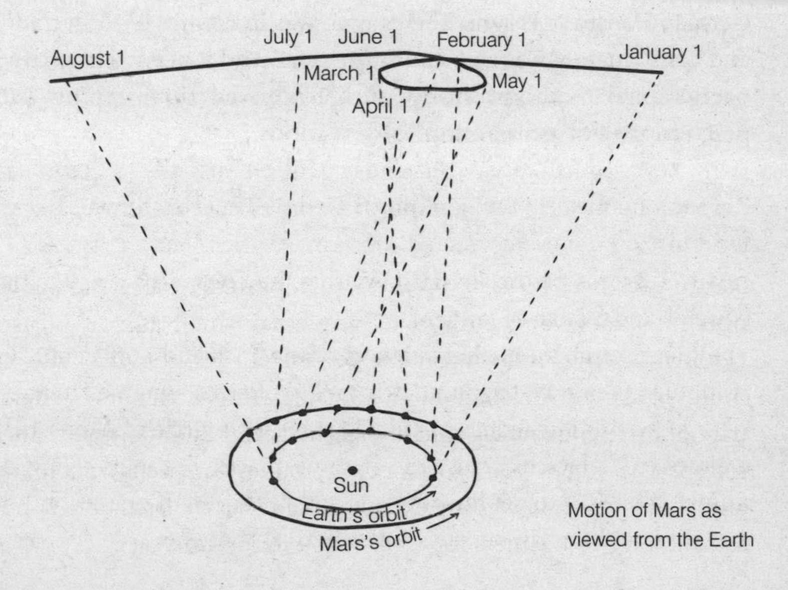

Motion of Mars as viewed from the Earth

Indian mathematicians generally used the half-chord, or *jya-ardha*, which is in effect the modern sine. Varahamihira calculated this function for 24 integer-multiples of 3°45′, up to 90°. Around 600, in the *Maha Bhaskariya*, Bhaskara gave a useful approximate formula for the sine of an acute angle, which he credited to Aryabhata. These authors derived a number of basic trigonometric formulas.

The Arabian mathematician Nasîr-Eddin's *Treatise on the Quadrilateral* combined plane and spherical geometry into a single unified development, and gave several basic formulas for spherical triangles. He treated the topic mathematically, rather than as a part of astronomy. But his work went unnoticed in the West until about 1450.

Because of the link with astronomy, almost all trigonometry was spherical until 1450. In particular, surveying – today a major user of trigonometry – was carried out using empirical methods, codified by the Romans. But in the mid-15th century, plane trigonometry began to come into its own, initially in the north German Hanseatic League. The League was in control of most trade, and consequently was rich and influential. And it needed improved navigational methods, along with improved timekeeping and practical uses of astronomical observations.

A key figure was Johannes Müller, usually known as Regiomontanus. He was a pupil of George Peuerbach, who began working on a new corrected version of the *Almagest*. In 1471, financed by his patron Bernard Walther, he computed a new table of sines and a table of tangents.

Other prominent mathematicians of the 15th and 16th centuries computed their own trigonometric tables, often to extreme accuracy. George Joachim Rhaeticus calculated sines for a circle of radius 10^{15} – effectively, tables accurate to 15 decimal places, but multiplying all numbers by 10^{15} to get integers – for all multiples of one second of arc. He stated the law of sines for spherical triangles

$$\frac{\sin a}{\sin A} = \frac{\sin b}{\sin B} = \frac{\sin c}{\sin C}$$

and the *law of cosines*

$$\cos a = \cos b \cos c + \sin b \sin c \cos A$$

in his *De Triangulis*, written in 1462–3 but not published until 1533. Here A, B, C are the angles of the triangle, and a, b, c are its sides – measured by the angles they determine at the centre of the sphere.

Vieta wrote widely on trigonometry, with his first book on the topic being the *Canon Mathematicus* of 1579. He collected and systematized various methods for solving triangles, that is, calculating all sides and angles from some subset of information. He invented new trigonometric identities, among them some interesting expressions for sines and cosines of integer multiples of θ in terms of the sine and cosine of θ.

Logarithms

The second theme of this chapter is one of the most important functions in mathematics: the logarithm, $\log x$. Initially, the logarithm was important because it satisfies the equation

$$\log xy = \log x + \log y$$

and can therefore be used to convert multiplications (which are cumbersome) into additions (which are simpler and quicker). To multiply two numbers x and y, first find their logarithms, add them and then find the number whose logarithm is that result (the *antilogarithm* of the result). This is the product xy.

Once mathematicians had calculated tables of logarithms, they could be used by anyone who understood the method. From the 17th century until the mid-20th, virtually all scientific calculations, especially astronomical ones, employed logarithms. From the 1960s onwards, however, electronic calculators and computers rendered logarithms obsolete for purposes of calculation. But the concept remained vital to mathematics, because logarithms had found fundamental roles in many parts of mathematics, including calculus and complex analysis. Also many physical and biological processes involve logarithmic behaviour.

Nowadays we approach logarithms by thinking of them as the reverse of exponentials. Using logarithms to base 10, which are a natural choice for decimal notation, we say that x is the logarithm of y if $y = 10^x$. For example, since $10^3 = 1000$, the logarithm of 1000 (to base 10) is 3. The basic property of logarithms follows from the exponential law

$$10^{a+b} = 10^a \times 10^b$$

However, in order for the logarithm to be useful, we have to be able to find a suitable x for any positive real y. Following the lead of Newton and others of his period, the main idea is that any rational power $10^{p/q}$ can be defined to be the qth root of 10^p. Since any real number x can be approximated arbitrarily closely by a rational number p/q, we can approximate 10^x by $10^{p/q}$. This is not the most efficient way to *calculate* the logarithm, but it is the simplest way to prove that it exists.

Historically, the discovery of logarithms was less direct. It began with John Napier, Baron of Murchiston in Scotland. He had a lifelong interest in efficient methods for calculation, and invented *Napier's rods* (or *Napier's bones*), a set of marked sticks that could be used to perform multiplication quickly and reliably by simulating pen-and-paper methods. Around 1594 he started working on a more theoretical method, and his writings tell us that it took him 20 years to perfect and publish it. It seems likely that he started with geometric progressions, sequences of numbers in which each term is obtained from the previous one by multiplying by a fixed number – such as the powers of 2

Plane Trigonometry

Nowadays trigonometry is first developed in the plane, where the geometry is simpler and the basic principles are easier to grasp. (It is curious how often new mathematical ideas are first developed in a complicated context, and the underlying simplicities emerge much later.) There is a law of sines, and a law of cosines, for plane triangles, and it is worth a quick digression to explain these. Consider a plane triangle with angles A, B, C and sides a, b, c.

Now the law of sines takes the form

$$\frac{a}{\sin A} = \frac{b}{\sin B} = \frac{c}{\sin C}$$

and the law of cosines is

$$a^2 = b^2 + c^2 - 2bc \cos A,$$

with companion formulas involving the other angles. We can use the law of cosines to find the angles of a triangle from its sides.

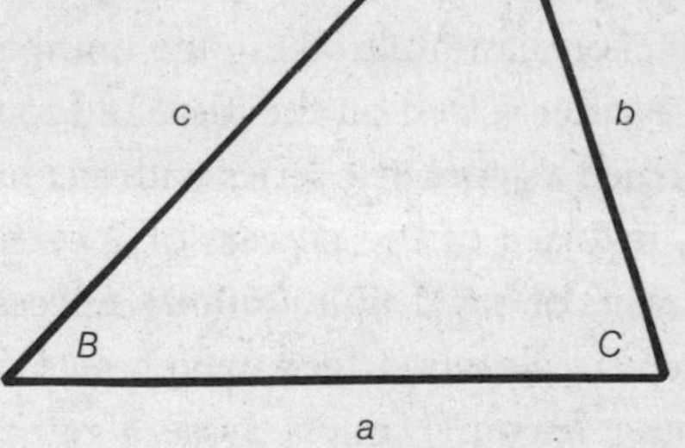

Sides and angles of a triangle

$$1 \quad 2 \quad 4 \quad 8 \quad 16 \quad 32 \quad \ldots$$

or powers of 10

$$1 \quad 10 \quad 100 \quad 1000 \quad 10{,}000 \quad 100{,}000 \quad \ldots$$

Here it had long been noticed that adding the exponents was equivalent to multiplying the powers. This was fine if you wanted to multiply two integer powers of 2, say, or two integer powers of 10. But there were big gaps between these numbers, and powers of 2 or 10 seemed not to help much when it came to problems like 57.681 × 29.443, say.

Napierian logarithms

While the good Baron was trying to somehow fill in the gaps in geometric progressions, the physician to King James VI of Scotland, James Craig, told Napier about a discovery that was in widespread use in Denmark, with the ungainly name *prosthapheiresis*. This referred to any process that converted products into sums. The main method in practical use was based on a formula discovered by Vieta:

$$\sin \frac{x+y}{2} \cos \frac{x-y}{2} = \frac{\sin x + \sin y}{2}$$

If you had tables of sines and cosines, you could use this formula to convert a product into a sum. It was messy, but it was still quicker than multiplying the numbers directly.

Napier seized on the idea, and found a major improvement. He formed a geometric series with a common ratio very close to 1. That is, in place of the powers of 2 or powers of 10, you should use powers of, say, 1.0000000001. Successive powers of such a number are very closely spaced, which gets rid of those annoying gaps. For some reason Napier chose a ratio slightly *less* than 1, namely 0.9999999. So his geometric sequence ran *backwards* from a large number to successively smaller ones. In fact, he started with 10,000,000 and then multiplied this by successive powers of 0.9999999. If we write Naplog x for Napier's logarithm of x, it has the curious feature that

$$\text{Naplog } 10{,}000{,}000 = 0$$

$$\text{Naplog } 9{,}999{,}999 = 1$$

and so on. The Napierian logarithm, Naplog x, satisfies the equation

$$\text{Naplog}(10^7 xy) = \text{Naplog}(x) + \text{Naplog}(y)$$

You can use this for calculation, because it is easy to multiply or divide by a power of 10, but it lacks elegance. It is, however, much better than Vieta's trigonometric formula.

Base ten logarithms

The next improvement came when Henry Briggs, the first Savilian professor of geometry at the University of Oxford, visited Napier. Briggs suggested replacing Napier's concept by a simpler one: the (base ten) logarithm, $L = \log_{10} x$, which satisfies the condition

$$x = 10^L.$$

Now

$$\log_{10} xy = \log_{10} x + \log_{10} y$$

and everything is easy. To find xy, add the logarithms of x and y and then find the antilogarithm of the result.

Before these ideas could be disseminated, Napier died; the year was 1617, and his description of his calculating rods, *Rhabdologia*, had just been published. His original method for calculating logarithms, the *Mirifici Logarithmorum Canonis Constructio*, appeared two years later. Briggs took up the task of computing a table of Briggsian (base 10, or *common*) logarithms. He did this by starting from $\log_{10} 10 = 1$ and taking successive square roots. In 1617 he published *Logarithmorum Chilias Prima*, the logarithms of the integers from 1 to 1000, stated to 14 decimal places. His 1624 *Arithmetic Logarithmica* tabulated common logarithms of numbers from 1 to 20,000 and from 90,000 to 100,000, also to 14 places.

What trigonometry did for them

Ptolemy's *Almagest* formed the basis of all studies of planetary motion prior to Johannes Kepler's discovery that orbits are elliptical. The observed movements of a planet are complicated by the relative motion of the Earth, which was not recognized in Ptolemy's time. Even if planets moved at uniform speed in circles, the Earth's motion round the Sun would effectively require a combination of two different circular motions, and an accurate model has to be distinctly more complicated than Ptolemy's model. Ptolemy's scheme of epicycles combines circular motions by making the centre of one circle revolve around another circle. This circle can itself revolve round a third circle, and so on. The geometry of uniform circular motion naturally involves trigonometric functions, and later astronomers used these for calculations of orbits.

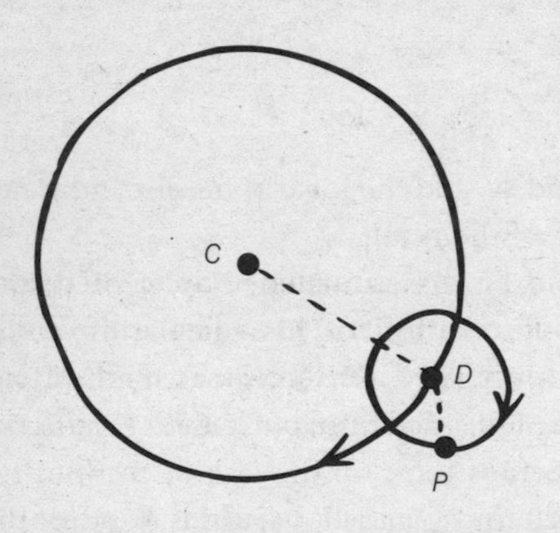

An epicycle. Planet *P* revolves uniformly around point *D*, which in turn revolves uniformly around point *C*.

The idea snowballed. John Speidell worked out logarithms of trigonometric functions (such as log sin *x*) published as *New Logarithmes* in 1619. The Swiss clockmaker Jobst Bürgi published his own work on logarithms in 1620, and may well have possessed the basic idea in 1588, well before Napier. But the historical development of mathematics depends on what people publish – in the original sense of make public – and ideas that remain private have no influence on anyone else. So credit, probably rightly, has to go to those people who put their ideas into print, or at least into widely circulated letters. (The exception is people who put the ideas of *others* into print without due credit. This is generally beyond the pale.)

The number *e*

Associated with Napier's version of logarithms is one of the most important numbers in mathematics, now denoted by the letter e. Its value is roughly 2.7128. It arises if we try to form logarithms by starting from a geometric series whose common ratio is very slightly larger than 1. This leads to the expression $(1 + 1/n)^n$, where n is a very large integer, and the larger n becomes, the closer this expression is to one special number, which we denote by e.

This formula suggests that there is a natural base for logarithms, and it is neither 10 nor 2, but e. The *natural logarithm* of x is whichever number y satisfies the condition $x = e^y$. In today's mathematics the natural logarithm is written $y = \log x$. Sometimes the base e is made explicit, as $y = \log_e x$, but this notation is mainly restricted to school mathematics, because in advanced mathematics and science the only logarithm of importance is the natural logarithm. Base ten logarithms are best for calculations in decimal notation, but natural logarithms are more fundamental mathematically.

The expression e^x is called the *exponential* of x, and it is one of the most important concepts in the whole of mathematics. The number e is one of those strange special numbers that appear in mathematics, and have major significance. Another such number is π. These two numbers are the tip of an iceberg – there are many others. They are also arguably the most important of the special numbers because they crop up all over the mathematical landscape.

What trigonometry does for us

Trigonometry is fundamental to surveying anything from building sites to continents. It is relatively easy to measure angles to high accuracy, but harder to measure distances, especially on difficult terrain. Surveyors therefore begin by making a careful measurement of one length, the *baseline*, that is, the distance between two specific locations. They then form a network of triangles, and use the measured angles, plus trigonometry, to calculate the sides of these triangles. In this manner, an accurate map of the entire area concerned can be constructed. This process is known as triangulation. To check its accuracy, a second distance measurement can be made once the triangulation is complete.

The figure here shows an early example, a famous survey carried out in South Africa in 1751 by the great astronomer Abbé Nicolas Louis de Lacaille. His main aim was to catalogue the stars of the southern skies, but to do this accurately he first had to measure the arc of a suitable line of longitude. To do this, he developed a triangulation to the north of Cape Town.

His result suggested that the curvature of the earth is less in southern latitudes than in northern ones, a surprising deduction which was verified by later measurements. The Earth is slightly pear-shaped. His cataloguing activities were so successful that he named 15 of the 88 constellations now recognized, having observed more than 10,000 stars using a small refracting telescope.

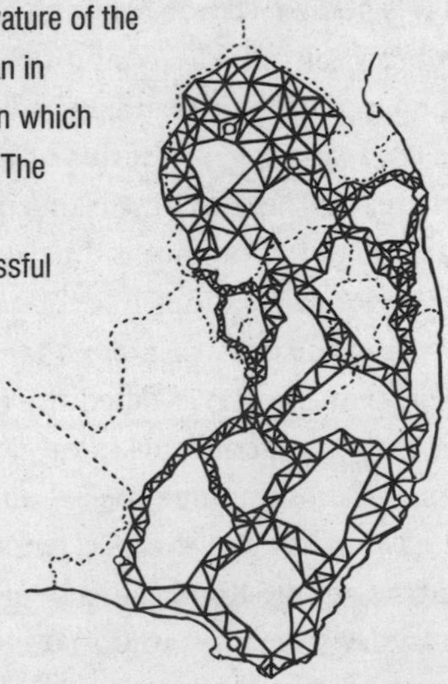

Lacaille's triangulation of South Africa

Where would we be without them?

It would be difficult to underestimate the debt that we owe to those far-seeing individuals who invented logarithms and trigonometry, and spent years calculating the first numerical tables. Their efforts paved the way to a quantitative scientific understanding of the natural world, and enabled worldwide travel and commerce by improving navigation and map-making. The basic techniques of surveying rely on trigonometric calculations. Even today, when surveying equipment uses lasers and the calculations are done on a custom-built electronic chip, the concepts that the laser and the chip embody are direct descendants of the trigonometry that intrigued the mathematicians of ancient India and Arabia.

Logarithms made it possible for scientists to do multiplication quickly and accurately. Twenty years of effort on a book of tables, by one mathematician, saved tens of thousands of man-years of work later on. It then became possible to carry out scientific analyses using pen and paper that would otherwise have been too time-consuming. Science could never have advanced without some such method. The benefits of such a simple idea have been incalculable.

CHAPTER 6

Curves and Coordinates

Geometry is algebra is geometry

Although it is usual to classify mathematics into separate areas, such as arithmetic, algebra, geometry and so on, this classification owes more to human convenience than the subject's true structure. In mathematics, there are no hard and fast boundaries between apparently distinct areas, and problems that seem to belong to one area may be solved using methods from another. In fact, the greatest breakthroughs often hinge upon making some unexpected connection between previously distinct topics.

Fermat

Greek mathematics has traces of such connections, with links between Pythagoras's Theorem and irrational numbers, and Archimedes's use of mechanical analogies to find the volume of the sphere. The true extent and influence of such cross-fertilization became undeniable in a short period ten years either side of 1630. During that brief time, two of the world's greatest mathematicians discovered a remarkable connection between algebra and geometry.

In fact, they showed that each of these areas can be converted into the other by using coordinates. Everything in Euclid, and the work of his successors, can be reduced to algebraic calculations. Conversely, everything in algebra can be interpreted in terms of the geometry of curves and surfaces.

It might seem that such connections render one or other area superfluous. If all geometry can be replaced by algebra, why do we need geometry? The answer is that each area has its own characteristic point of view, which can on occasion be very penetrating and powerful. Sometimes it is best to think geometrically, and sometimes algebraic thinking is superior.

The first person to describe coordinates was Pierre de Fermat. Fermat is best known for his work in number theory, but he also studied many other areas of mathematics, including probability, geometry and applications to optics. Around 1620, Fermat was trying to understand the geometry of curves, and he started by reconstructing, from what little information was available, a lost book by Apollonius called *On Plane Loci*. Having done this, Fermat embarked upon his own investigations, writing them up in 1629 but not publishing them until 50 years later, as *Introduction to Plane and Solid Loci*. By so doing, he discovered the advantages of rephrasing geometrical concepts in algebraic terms.

Locus, plural *loci*, is an obsolete term today, but it was common even in 1960. It arises when we seek all points in the plane or space

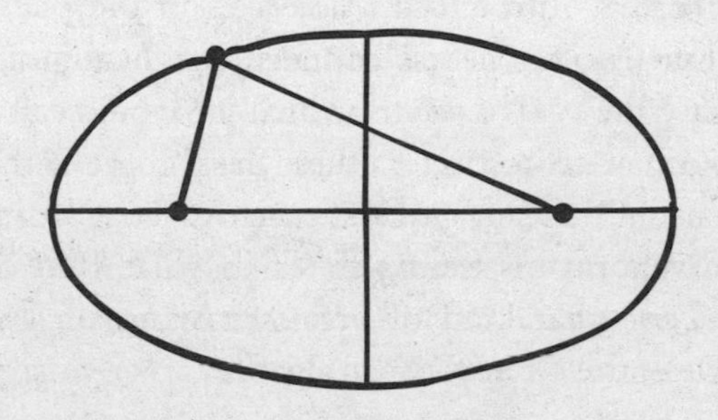

Focal property of the ellipse

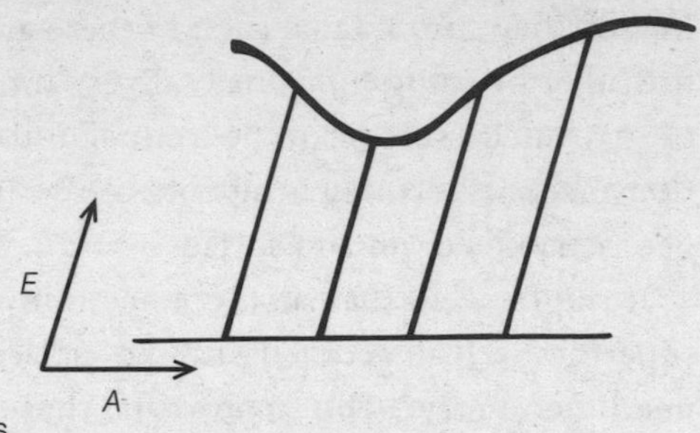

Fermat's approach to coordinates

that satisfy particular geometric conditions. For example, we might ask for the locus of all points whose distances from two other fixed points always add up to the same total. This locus turns out to be an ellipse with the two points as its foci. This property of the ellipse was known to the Greeks.

Fermat noticed a general principle: if the conditions imposed on the point can be expressed as a single equation involving two unknowns, the corresponding locus is a curve – or a straight line, which we consider to be a special kind of curve to avoid making needless distinctions. He illustrated this principle by a diagram in which the two unknown quantities *A* and E are represented as distances in two distinct directions.

He then listed some special types of equation connecting *A* and E, and explained what curves they represent. For instance, if $A^2 = 1 + E^2$ then the locus concerned is a hyperbola.

In modern terms, Fermat introduced *oblique* axes in the plane (oblique meaning that they do not necessarily cross at right angles). The variables *A* and E are the two *coordinates*, which we would call *x* and *y*, of any given point with respect to these axes. So Fermat's principle effectively states that any equation in two coordinate variables defines a curve, and his examples tell us what kind of equations correspond to what kind of curve, drawing on the standard curves known to the Greeks.

Descartes

The modern notion of coordinates came to fruition in the work of Descartes. In everyday life, we are familiar with spaces of two and three dimensions, and it takes a major effort of imagination for us to contemplate other possibilities. Our visual system presents the outside world to each eye as a two-dimensional image – like the picture on a TV screen. Slightly different images from each eye are combined by the brain to provide a sense of depth, through which we perceive the surrounding world as having three dimensions.

The key to multidimensional spaces is the idea of a coordinate system, which was introduced by Descartes in an appendix *La Géometrie* to his book *Discours de la Méthode*. His idea is that the geometry of the plane can be reinterpreted in algebraic terms. His approach is essentially the same as Fermat's. Choose some point in the plane and call it the origin. Draw two axes, lines that pass through the origin and meet at right angles. Label one axis with the symbol x and the other with the symbol y. Then any point P in the plane is determined by the pair of distances (x, y), which tells us how far the point is from the origin when measured parallel to the x- and y-axes, respectively.

For example, on a map, x might be the distance east of the origin (with negative numbers representing distances to the west), whereas y might be the distance north of the origin (with negative numbers representing distances to the south).

Coordinates work in three-dimensional space too, but now two numbers are not sufficient to locate a point. However, three numbers are. As well as the distances east–west and north–south, we need to know how far the point is above or below the origin. Usually we use a positive number for distances above, and a negative one for distances below. Coordinates in space take the form (x, y, z).

This is why the plane is said to be two-dimensional, whereas space is three-dimensional. The *number of dimensions* is given by how many numbers we need to specify a point.

René Descartes

1596–1650

Descartes first began to study mathematics in 1618, as a student of the Dutch scientist Isaac Beeckman. He left Holland to travel through Europe, and joined the Bavarian army in 1619. He continued to travel between 1620 and 1628, visiting Bohemia, Hungary, Germany, Holland, France and Italy. He met Mersenne in Paris in 1622, and corresponded regularly with him thereafter, which kept him in touch with most of the leading scholars of the period.

In 1628 Descartes settled in Holland, and started his first book *Le Monde, ou Traité de la Lumière*, on the physics of light. Publication was delayed when Descartes heard of Galileo Galilei's house arrest, and Descartes got cold feet. The book was published, in an incomplete form, after his death. Instead, he developed his ideas on logical thinking into a major work published in 1637: *Discours de la Méthode*. The book had three appendices: *La Dioptrique*, *Les Météores* and *La Géométrie*.

His most ambitious book, *Principia Philosophiae*, was published in 1644. It was divided into four parts: *Principles of Human Knowledge*, *Principles of Material Things*, *The Visible World* and *The Earth*. It was an attempt to provide a unified mathematical foundation for the entire physical universe, reducing everything in nature to mechanics.

In 1649 Descartes went to Sweden, to act as tutor to Queen Christina. The Queen was an early riser, whereas Descartes habitually rose at 11 o'clock. Tutoring the Queen in mathematics at 5 o'clock every morning, in a cold climate, imposed a considerable strain on Descartes's health. After a few months, he died of pneumonia.

Coordinates as Used Today

The early development of coordinate geometry will make more sense if we first explain how the modern version goes. There are several variants, but the commonest begins by drawing two mutually perpendicular lines in the plane, called *axes*. Their common meeting point is the *origin*. The axes are conventionally arranged so that one of them is horizontal and the other vertical.

Along both axes we write the whole numbers, with negative numbers going in one direction and positive numbers in the other. Conventionally, the horizontal axis is called the x-axis, and the vertical one is the y-axis. The symbols x and y are used to represent points on those respective axes – distances from the origin. A general point in the plane, distance x along the horizontal axis and distance y along the vertical one, is labelled with a pair of numbers (x, y). These numbers are the *coordinates* of that point.

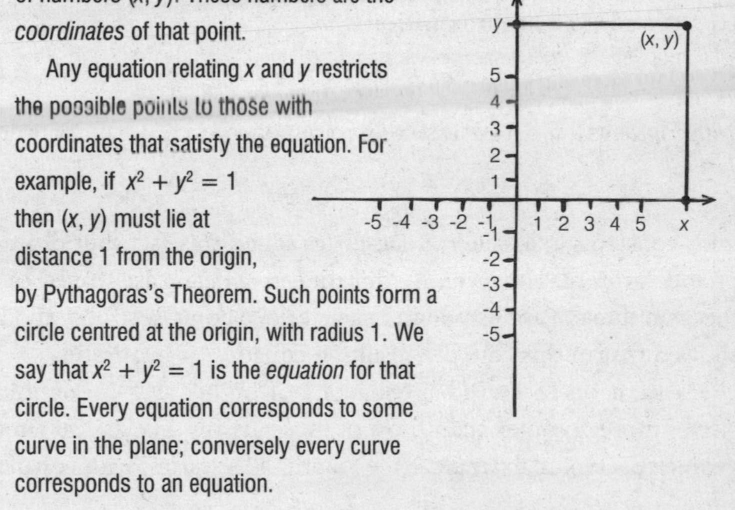

Any equation relating x and y restricts the possible points to those with coordinates that satisfy the equation. For example, if $x^2 + y^2 = 1$ then (x, y) must lie at distance 1 from the origin, by Pythagoras's Theorem. Such points form a circle centred at the origin, with radius 1. We say that $x^2 + y^2 = 1$ is the *equation* for that circle. Every equation corresponds to some curve in the plane; conversely every curve corresponds to an equation.

In three-dimensional space, a single equation involving x, y and z usually defines a surface. For example, $x^2 + y^2 + z^2 = 1$ states that the point (x, y, z) is always a distance 1 unit from the origin, which implies that it lies on the unit sphere whose centre is the origin.

Notice that the word 'dimension' is not actually defined here in its own right. We do not find the number of dimensions of a space by finding some things called dimensions and then counting them. Instead, we work out *how many numbers* are needed to specify where a location in the space is, and that is the number of dimensions.

Cartesian coordinates

Cartesian coordinate geometry reveals an algebraic unity behind the conic sections – curves that the Greeks had constructed as sections of a double cone. Algebraically, it turns out that the conic sections are the next simplest curves after straight lines. A straight line corresponds to a linear equation

$$ax + by + c = 0$$

with constants a, b, c. A conic section corresponds to a quadratic equation

$$ax^2 + bxy + cy^2 + dx + ey + f = 0$$

with constants a, b, c, d, e, f. Descartes stated this fact, but did not provide a proof. However, he did study a special case, based on a theorem due to Pappus which characterized conic sections, and he showed that in this case the resulting equation is quadratic.

He went on to consider equations of higher degree, defining curves more complex than most of those arising in classical Greek geometry. A typical example is the folium of Descartes, with equation

$$x^3 + y^3 - 3axy = 0$$

which forms a loop with two ends that tend to infinity.

Perhaps the most important contribution made by the concept of coordinates occurred here: Descartes moved away from the

Greek view of curves as things that are constructed by specific geometric means, and saw them as the visual aspect of any algebraic formula. As Isaac Newton remarked in 1707, 'The moderns advancing yet much further [than the Greeks] have received into geometry all lines that can be expressed by equations'.

Later scholars invented numerous variations on the Cartesian coordinate system. In a letter of 1643 Fermat took up Descartes's ideas and extended them to three dimensions. Here he mentioned surfaces such as ellipsoids and paraboloids, which are determined by quadratic equations in the three variables x, y, z. An influential contribution was the introduction of *polar* coordinates by Jakob Bernoulli in 1691. He used an angle θ and a distance r to determine points in the plane instead of a pair of axes. Now the coordinates are (r, θ).

Again, equations in these variables specify curves. But now, simple equations can specify curves that would become very complicated in Cartesian coordinates. For example the equation $r = \theta$ corresponds to a spiral, of the kind known as an *Archimedean spiral*.

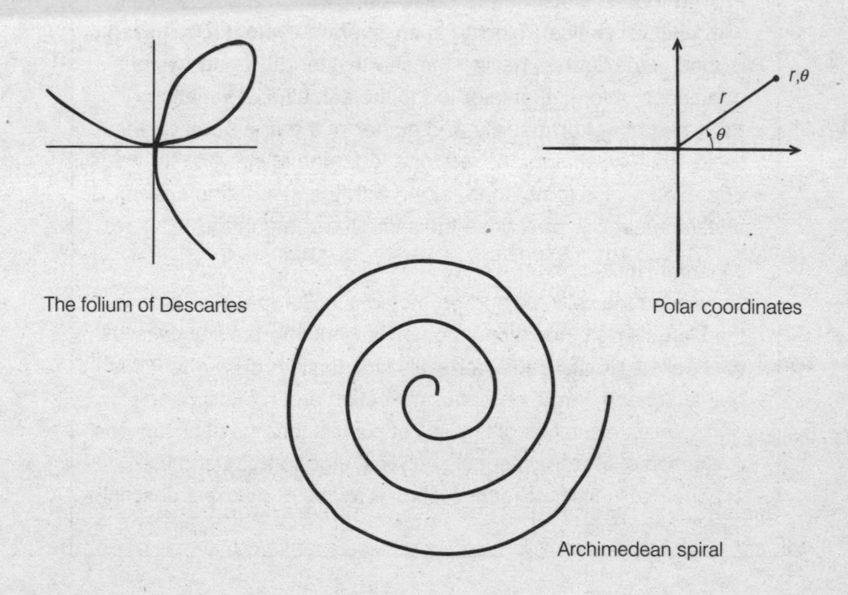

The folium of Descartes

Polar coordinates

Archimedean spiral

Which Bernoulli did What?

A Bernoulli Checklist

The Swiss Bernoulli family had a huge influence on the development of mathematics. Over some four generations they produced significant mathematics, both pure and applied. Often described as a mathematical mafia, the Bernoullis typically started out in careers like law, medicine or the church, but eventually reverted to type and became mathematicians, either professional or amateur.

Many different mathematical concepts bear the name Bernoulli. This is not always the *same* Bernoulli. Rather than providing biographical details about them, here is a summary of who did what.

Jacob I (1654–1705)

Polar coordinates, formula for the radius of curvature of a plane curve. Special curves, such as the catenary and lemniscate. Proved that an isochrone (a curve along which a body will fall with uniform vertical velocity) is an inverted cycloid. Discussed isoperimetric figures, having the shortest length under various conditions, a topic that later led to the calculus of variations. Early student of probability and author of the first book on the topic, *Ars Conjectandi*. Asked for a logarithmic spiral to be engraved on his tombstone, along with the inscription *Eadem mutata resurgo* (I shall arise the same though changed).

Johann I (1667–1748)

Developed the calculus and promoted it in Europe. His student the Marquis de L'Hôpital put Johann's work into the first calculus textbook. 'L'Hôpital's rule' for evaluating limits reducing to 0/0 is due to Johann. Wrote on optics (reflection and refraction), orthogonal trajectories of families of curves, lengths of curves and evaluation of areas by series, analytical trigonometry and the exponential function. Brachistochrone (curve of quickest descent), length of cycloid.

Nicolaus I (1687–1759)
Held Galileo's chair of mathematics at Padua. Wrote on geometry and differential equations. Later, taught logic and law. A gifted but not very productive mathematician. Corresponded with Leibniz, Euler and others – his main achievements are scattered among 560 items of correspondence. Formulated the St Petersburg Paradox in probability.

Criticized Euler's indiscriminate use of divergent series. Assisted in publication of Jacob Bernoulli's *Ars Conjectandi*. Supported Leibniz in his controversy with Newton.

Nicolaus II (1695–1726)
Called to the St. Petersburg Academy, died by drowning eight months later. Discussed the St Petersburg Paradox with Daniel.

Daniel (1700–1782)
Most famous of Johann's three sons. Worked on probability, astronomy, physics and hydrodynamics. His *Hydrodynamica* of 1738 contains *Bernoulli's principle* about the relation between pressure and velocity. Wrote on tides, the kinetic theory of gases and vibrating strings. Pioneer in partial differential equations.

Johann II (1710–1790)
Youngest of Johann's three sons. Studied law but became professor of mathematics at Basel. Worked on the mathematical theory of heat and light.

Johann III (1744–1807)
Like his father, studied law but then turned to mathematics. Called to the Berlin Academy at age 19. Wrote on astronomy, chance and recurring decimals.

Jacob II (1759–1789)
Important works on elasticity, hydrostatics, and ballistics.

Functions

An important application of coordinates in mathematics is a method to represent functions graphically.

A **function** is not a number, but a recipe that starts from some number and calculates an associated number. The recipe involved is often stated as a formula, which assigns to each number, x (possibly in some limited range), another number, $f(x)$.

For example, the square root function is defined by the rule $f(x) = \sqrt{x}$, that is, take the square root of the given number. This recipe requires x to be positive. Similarly the square function is defined by $f(x) = x^2$, and this time there is no restriction on x.

We can picture a function geometrically by defining the y-coordinate, for a given value of x, by $y = f(x)$. This equation states a relationship between the two coordinates, and therefore determines a curve. This curve is called the **graph** of the function f.

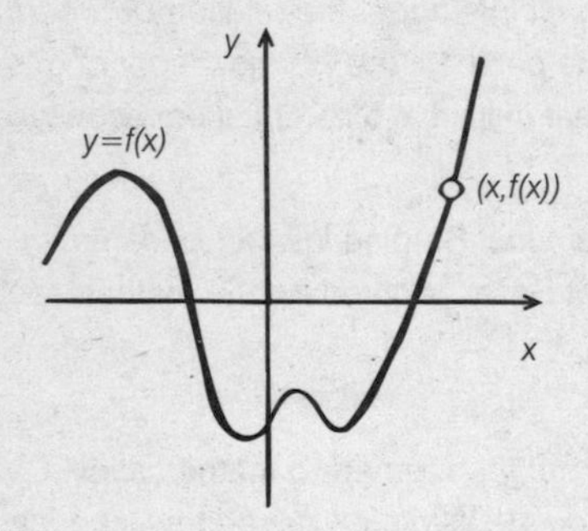

Graph of a function *f*

The graph of the function $f(x) = x^2$ turns out to be a parabola. That of the square root $f(x) = \sqrt{x}$ is half a parabola, but lying on its side. More complicated functions lead to more complicated curves. The graph of the sine function $y = \sin x$ is a wiggly wave.

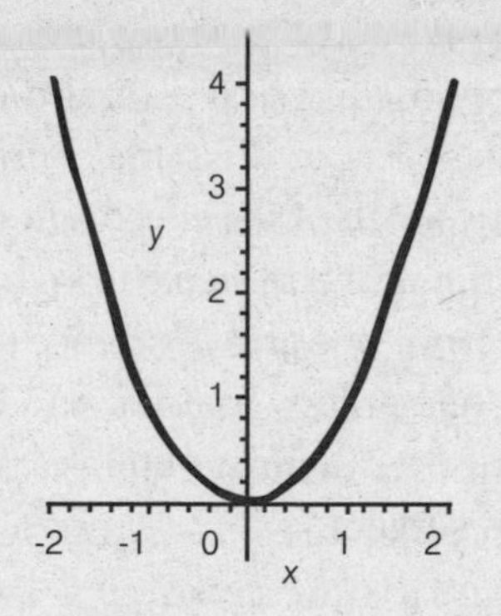

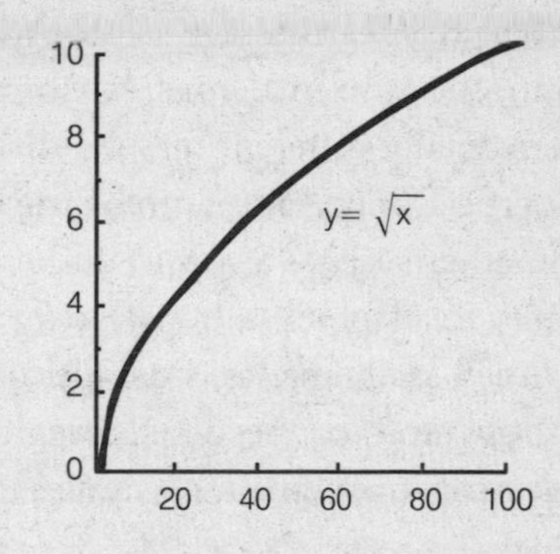

Graphs of the square and square root

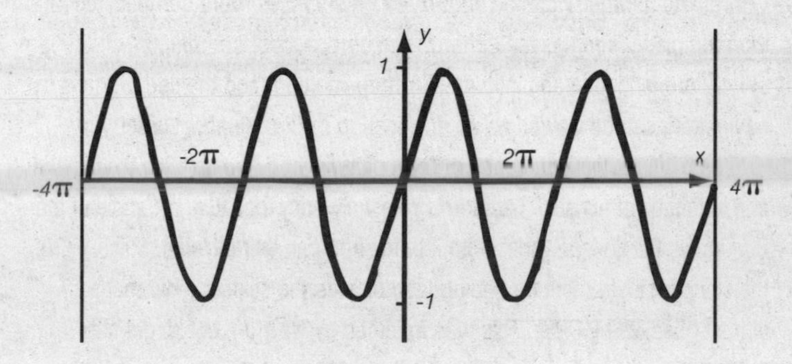

Graph of the sine function

Coordinate geometry today

Coordinates are one of those simple ideas that has had a marked influence on everyday life. We use them everywhere, usually without noticing what we're doing. Virtually all computer graphics employ an internal coordinate system, and the geometry that appears on the screen is dealt with algebraically. An operation as simple as rotating a digital photograph through a few degrees, to get the horizon horizontal, relies on coordinate geometry.

The deeper message of coordinate geometry is about cross-connections in mathematics. Concepts whose physical realizations seem totally different may be different aspects of the same thing. Superficial appearances can be misleading. Much of the effectiveness of mathematics as a way to understand the universe stems from its ability to adapt ideas, transferring them from one area of science to another. Mathematics is the ultimate in technology transfer. And it is those cross-connections, revealed to us over the past 4000 years, that make mathematics a single, unified subject.

What coordinates did for them

Coordinate geometry can be employed on surfaces more complicated than the plane, such as the sphere. The commonest coordinates on the sphere are longitude and latitude. So map-making, and the use of maps in navigation, can be viewed as an application of coordinate geometry.

The main navigational problem for a captain was to determine the latitude and longitude of his ship. Latitude is relatively easy, because the angle of the Sun above the horizon varies with latitude and can be tabulated. Since 1730, the standard instrument for finding latitude was the sextant (now made almost obsolete by GPS). This was invented by Newton, but he did not publish it. It was independently rediscovered by the English mathematician John Hadley and the American inventor Thomas Godfrey. Previous navigators had used the astrolabe, which goes back to medieval Arabia. Longitude is trickier. The problem was eventually solved by constructing a highly accurate clock, which was set to local time at the start of the voyage. The time of sunrise and sunset, and the movements of the Moon and stars, depend on longitude, making it possible to determine longitude by comparing local time with that on the clock. The story of John Harrison's invention of the chronometer, which solved the problem, is famously told in Dava Sobel's *Longitude*.

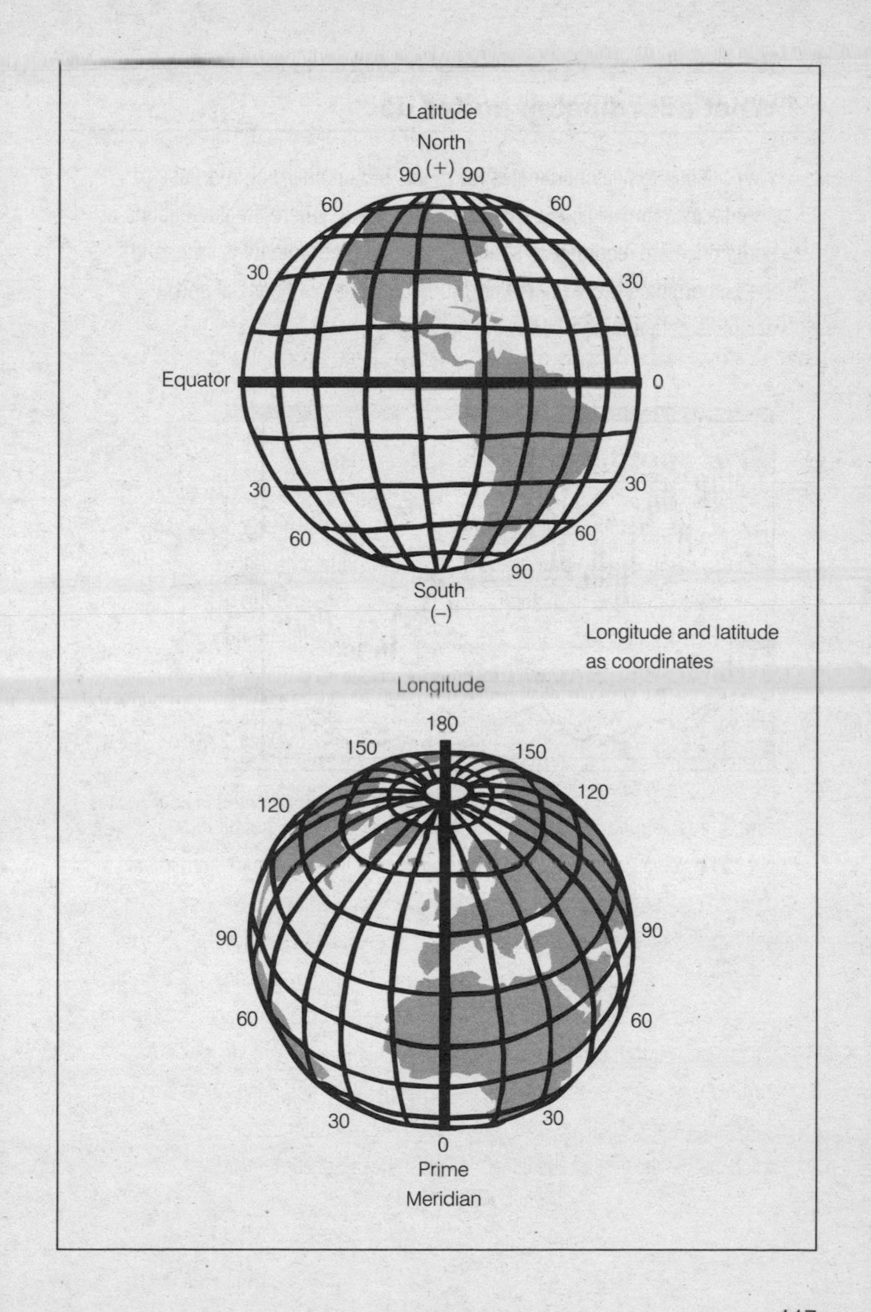

Longitude and latitude as coordinates

What coordinates do for us

We continue to use coordinates for maps, but another common use of coordinate geometry occurs in the stock market, where the fluctuations of some price are recorded as a curve. Here the x-coordinate is time, and the y-coordinate is the price. Enormous quantities of financial and scientific data are recorded in the same way.

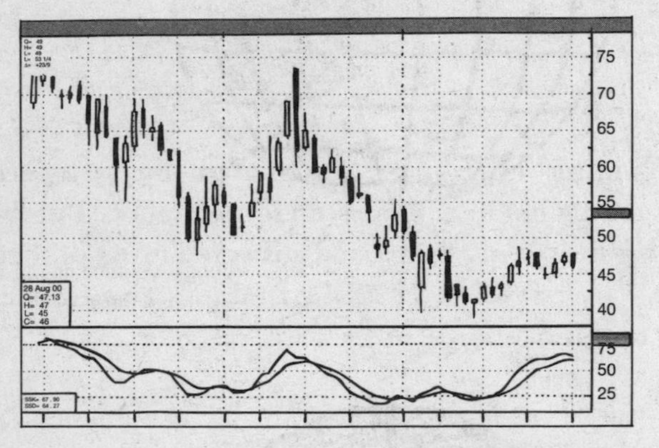

Stock market data represented in coordinates

CHAPTER 7

Patterns in Numbers

The origins of number theory

Despite becoming ever more fascinated by geometry, mathematicians did not lose their interest in numbers. But they did start asking deeper questions, and answered many of them. A few had to wait for more powerful techniques. Some remain unanswered to this day.

Number theory

There is something fascinating about numbers. Plain, unadorned whole numbers, 1, 2, 3, 4, 5, ... What could possibly be simpler? But that simple exterior conceals hidden depths, and many of the most baffling questions in mathematics are about apparently straightforward properties of whole numbers. This area is known as *number theory*, and it turns out to be difficult precisely because its ingredients are so basic. The very simplicity of whole numbers leaves little scope for clever techniques.

The earliest serious contributions to number theory – that is, complete with proofs, not just assertions – are found in the works of Euclid, where the ideas are thinly disguised as geometry. The subject was developed into a distinct area of mathematics by the Greek, Diophantus, some of whose writings survive as later copies. Number theory was given a big boost in the 1600s by Fermat, and developed by Leonhard Euler, Joseph-Louis Lagrange and Carl Friedrich Gauss into a deep and extensive branch of mathematics which touched upon many other areas, often seemingly unrelated. By the end of the 20th century these connections had been used to answer some – though not all – of the ancient puzzles, including a very famous conjecture made by Fermat around 1650, known as his Last Theorem.

For most of its history, number theory has been about the internal workings of mathematics itself, with few connections to the real world. If ever there was a branch of mathematical thought that lived in the heady reaches of the ivory towers, it was number theory. But the advent of the digital computer has changed all that. Computers work with electronic representations of whole numbers, and the problems and opportunities raised by computers frequently lead to number theory. After 2500 years as a purely intellectual exercise, number theory has finally made an impact on everyday life.

Primes

Anyone who contemplates the multiplication of whole numbers eventually notices a fundamental distinction.

Many numbers can be broken up into smaller pieces, in the sense that the number concerned arises by multiplying those pieces together. For instance, 10 is 2 × 5, and 12 is 3 × 4. Some numbers, however, do not break up in this manner. There is no way to express 11 as the product of two *smaller* whole numbers; the same goes for 2, 3, 5, 7 and many others.

The numbers that can be expressed as the product of two smaller numbers are said to be *composite*. Those that cannot be so expressed are *prime*. According to this definition, the number 1 should be considered prime, but for good reasons it is placed in a special class of its own and called a *unit*. So the first few primes are the numbers

2 3 5 7 11 13 17 19 23 29 31 37 41

As this list suggests, there is no obvious pattern to the primes (except that all but the first are odd). In fact, they seem to occur somewhat irregularly, and there is no simple way to predict the next number on the list. Even so, there is no question that this number is somehow *determined* – just test successive numbers until you find the next prime.

Despite, or perhaps because of, their irregular distribution, primes are of vital importance in mathematics. They form the basic building blocks for all numbers, in the sense that larger numbers are created by multiplying smaller ones. Chemistry tells us that any molecule, however complicated, is built from atoms – chemically indivisible particles of matter. Analogously, mathematics tells us that any number, however big it may be, is built from primes – indivisible numbers. So primes are the atoms of number theory.

This feature of primes is useful because many questions in mathematics can be solved for all whole numbers provided they can be solved for the primes, and primes have special properties that sometimes make the solution of the question easier. This dual aspect of the primes – important but ill-behaved – excites the mathematician's curiosity.

Euclid

Euclid introduced primes in Book VII of the *Elements*, and he gave proofs of three key properties. In modern terminology, these are:

(i) Every number can be expressed as a product of primes.
(ii) This expression is unique except for the order in which the primes occur.
(iii) There are infinitely many primes.

What Euclid actually stated and proved is slightly different. Proposition 31, Book VII tells us that any composite number is measured by some prime – that is, it can be divided exactly by that prime. For example, 30 is composite, and it is exactly divisible by

Why Uniqueness of Prime Factors is not Obvious

Since the primes are the atoms of number theory, it might seem obvious that the *same* atoms always turn up when a number is broken into primes. After all, atoms are the indivisible pieces. If you could break a number up in two distinct ways, wouldn't that involve splitting an atom? But here the analogy with chemistry is slightly misleading.

To see that uniqueness of prime factorization is *not* obvious, we can work with a restricted set of numbers:

1 5 9 13 17 21 25 29

and so on. These are the numbers that are one greater than a multiple of 4. Products of such numbers also have the same property, so we can build such numbers up by multiplying smaller numbers of the same type. Define a 'quasiprime' to be any number in this list that is not the product of two smaller numbers *in the list*. For instance, 9 is quasiprime: the only smaller numbers in the list are 1 and 5, and their product is not 9. (It is still true that $9 = 3 \times 3$, of course, but the number 3 is not in the list.)

It is obvious – and true – that every number in the list is a product of quasiprimes. However, although these quasiprimes are the atoms of the set, something rather strange happens. The number 693 breaks up in two different ways: $693 = 9 \times 77 = 21 \times 33$, and all four factors, 9, 21, 33 and 77, are *quasiprime*. So uniqueness of factorization fails for this type of number.

several primes, among them 5 – in fact, $30 = 6 \times 5$. By repeating this process of pulling out a prime divisor, or *factor*, we can break any number down into a product of primes. Thus, starting from $30 = 6 \times 5$, we observe that 6 is also composite, with $6 = 2 \times 3$. Now $30 = 2 \times 3 \times 5$, and all three factors are prime.

If instead we had started from $30 = 10 \times 3$, then we would break down 10 instead, as $10 = 2 \times 5$, leading to $30 = 2 \times 5 \times 3$. The same three primes occur, but multiplied in a different order – which of course does not affect the result. It may seem obvious that however we break a number into primes, we always get the same result except for order, but this turns out to be tricky to prove. In fact, similar statements in some related systems of numbers turn out to be *false*, but for ordinary whole numbers the statement is true. Prime factorization is *unique*. Euclid proves the key fact needed to establish uniqueness in Proposition 30, Book VII of the *Elements*: if a prime divides the product of two numbers, then it must divide at least one of those numbers. Once we know Proposition 30, the uniqueness of prime factorization is a straightforward consequence.

Proposition 20, Book IX states that: 'Prime numbers are more than any assigned multitude of prime numbers.' In modern terms, this

The Largest Known Prime

There is no largest prime, but the largest *known* prime as of May 2009 is $2^{43,112,609} - 1$, which has 12,978,189 decimal digits. Numbers of the form $2^p - 1$, with p prime, are called Mersenne primes, because Mersenne conjectured in his *Cogitata Physica-Mathematica* of 1644 that these numbers are prime for $p = 2, 3, 5, 7, 13, 17, 19, 31, 67, 127$ and 257, and composite for all other whole numbers up to 257.

There are special high-speed methods for testing such numbers to see if they are primes, and we now know that Mersenne made five mistakes. His numbers are composite when $p = 67$ and 257, and there are three more primes with $p = 61, 89, 107$. Currently 44 Mersenne primes are known. Finding new ones is a good way to test new supercomputers, but has no practical significance.

means that the list of primes is infinite. The proof is given in a representative case: suppose that there are only three prime numbers, a, b and c. Multiply them together and add one, to obtain $abc + 1$. This number must be divisible by some prime, but that prime cannot be any of the original three, since these divide abc exactly, so they cannot also divide $abc + 1$, since they would then divide the difference, which is 1. We have therefore found a new prime, contradicting the assumption that a, b, c are all the primes there are.

Although Euclid's proof employs three primes, the same idea works for a longer list. Multiply all primes in the list, add one and then take some prime factor of the result; this always generates a prime that is not on the list. Therefore no finite list of primes can ever be complete.

Diophantus

We have mentioned Diophantus of Alexandria in connection with algebraic notation, but his greatest influence was in number theory. Diophantus studied general questions, rather than specific numerical ones, although his *answers* were specific numbers. For example: 'Find three numbers such that their sum, and the sum of any two, is a perfect square.' His answer is 41, 80 and 320. To check: the sum of all three is $441 = 21^2$. The sums of pairs are $41 + 80 = 11^2$, $41 + 320 = 19^2$ and $80 + 320 = 20^2$.

One of the best known equations solved by Diophantus is a curious offshoot of Pythagoras's Theorem. We can state the theorem algebraically: if a right triangle has sides a, b, c with c being the longest, then $a^2 + b^2 = c^2$. There are some special right triangles for which the sides are whole numbers. The simplest and best known is when a, b, c are 3, 4, 5, respectively; here $3^2 + 4^2 = 9 + 16 = 25 = 5^2$. Another example, the next simplest, is $5^2 + 12^2 = 13^2$.

In fact, there are infinitely many of these *Pythagorean triples*. Diophantus found all possible whole number solutions of what we now write as the equation $a^2 + b^2 = c^2$. His recipe is to take any two whole numbers, and form the difference between their squares,

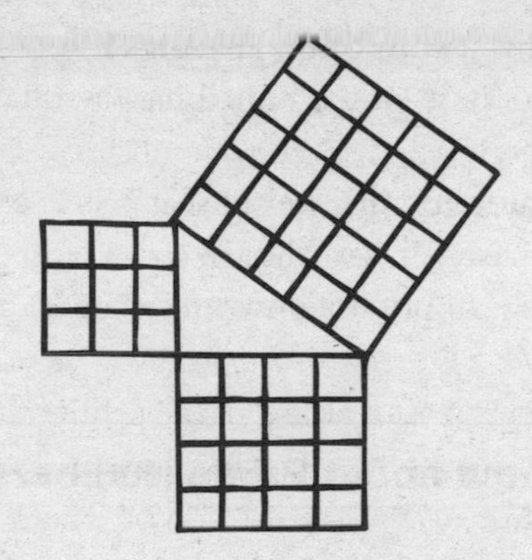

The 3-4-5 right-angled triangle

twice their product and the sum of their squares. These three numbers always form a Pythagorean triple, and all such triangles arise in this manner provided we also allow all three numbers to be multiplied by some constant. If the numbers are 1 and 2, for example, we get the famous 3-4-5 triangle. In particular, since there are infinitely many ways to choose the two numbers, there exist infinitely many Pythagorean triples.

Fermat

After Diophantus, number theory stagnated for over a thousand years, until it was taken up by Fermat, who made many important discoveries. One of his most elegant theorems tells us exactly when a given integer n is a sum of two perfect squares: $n = a^2 + b^2$. The solution is simplest when n is prime. Fermat observed that there are three basic types of prime:

(i) The number 2, the only even prime.
(ii) Primes that are 1 greater than a multiple of 4, such as 5, 13, 17 and so on – these primes are all odd.
(iii) Primes that are 1 less than a multiple of 4, such as 3, 7, 11 and so on – these primes are also odd.

He proved that a prime is a sum of two squares if it belongs to categories (i) or (ii), and it is not a sum of two squares if it belongs to category (iii).

For instance, 37 is in category (ii), being $4 \times 9 + 1$, and $37 = 6^2 + 1^2$, a sum of two squares. In contrast, $31 = 4 \times 8 - 1$ is in category (iii), and if you try all possible ways to write 31 as a sum

What We Don't Know about Prime Numbers

Even today, the primes still have some secrets. Two famous unsolved problems are the Goldbach Conjecture and the Twin Primes Conjecture.

Christian Goldbach was an amateur mathematician who corresponded regularly with Euler. In a letter of 1742, he described evidence that every whole number greater than 2 is the sum of three primes. Goldbach viewed 1 as a prime, which we no longer do; as a consequence, we would now exclude the numbers $3 = 1 + 1 + 1$ and $4 = 2 + 1 + 1$. Euler proposed a stronger conjecture: that every even number greater than 2 is the sum of two primes. For example, $4 = 2 + 2$, $6 = 3 + 3$, $8 = 5 + 3$, $10 = 5 + 5$ and so on. This conjecture implies Goldbach's. Euler was confident that his conjecture was true, but could not find a proof, and the conjecture is still open. Computer experiments have shown it to be true for every even number up 10^{18}. The best known result was obtained by Chen Jing-Run in 1973 using complicated techniques from analysis. He proved that every sufficiently large even number is the sum of two primes, or a prime and an almost-prime (the product of two primes).

The twin prime conjecture is much older, and goes back to Euclid. It states that there are infinitely many *twin primes* p and $p + 2$. Examples of twin primes are 5, 7 and 11, 13. Again, no proof or disproof is known. In 1966, Chen proved that there are infinitely many primes p, such that $p + 2$ is either prime or almost-prime. Currently, the largest known twin primes are $2{,}003{,}663{,}613 \times 2^{195{,}000} \pm 1$, found by Eric Vautier, Patrick McKibbon and Dmitri Gribenko in 2007.

of two squares, you will find that nothing works. (For instance, 31 = 25 + 6, where 25 is a square, but 6 is not.)

The upshot is that a number is a sum of two squares if and only if every prime divisor of the form $4k - 1$ occurs to an even power. Using similar methods, Joseph-Louis Lagrange proved in 1770 that every positive integer is a sum of four perfect squares (including one or more 0s if necessary). Fermat had previously stated this result, but no proof is recorded.

One of Fermat's most influential discoveries is also one of the simplest. It is known as Fermat's Little Theorem, to avoid confusion with his Last (sometimes called Great) Theorem, and it states that if p is any prime and a is any whole number, then $a^p - a$ is a multiple of p. The corresponding property is *usually* false when p is composite, but not always.

Fermat's most celebrated result took 350 years to prove. He stated it around 1640, and he claimed a proof, but all we know of his work is a short note. Fermat owned a copy of Diophantus's *Arithmetica*, which inspired many of his investigations, and he often wrote down his own ideas in the margin. At some point he must have been thinking about the Pythagorean equation: add two squares to get a square. He wondered what would happen if instead of squares you tried cubes, but found no solutions. The same problem arose for fourth, fifth or higher powers.

In 1670 Fermat's son Samuel published an edition of Bachet's translation of the *Arithmetica*, which included Fermat's marginal notes. One such note became notorious: the statement that if $n \geq 3$, the sum of two nth powers is never an nth power. The marginal note states 'To resolve a cube into the sum of two cubes, a fourth power into two fourth powers or, in general, any power higher than the second into two of the same kind is impossible; of which fact I have found a remarkable proof. The margin is too small to contain it.'

It seems unlikely that his proof, if it existed, was correct. The first,

Pierre de Fermat

1601–1665

Pierre Fermat was born at Beaumont-de-Lomagne in France in 1601, son of the leather merchant Dominique Fermat and Claire de Long, the daughter of a family of lawyers. By 1629 he had made important discoveries in geometry and the forerunner of the calculus, but he chose law as a career, becoming a Councillor at the parliament of Toulouse in 1631. This entitled him to add the 'de' to his name. As an outbreak of plague killed off his superiors, he advanced rapidly. In 1648 he became a King's Councillor in the local parliament of Toulouse, where he served for the rest of his life, reaching the most senior level of the Criminal Court in 1652.

He never held an academic position, but mathematics was his passion. In 1653 he contracted plague, was rumoured to be dead, but survived. He carried out extensive correspondence with other intellectuals, notably the mathematician Pierre de Carcavi and the monk Marin Mersenne.

He worked in mechanics, optics, probability and geometry, and his method for locating the maximum and minimum values of a function paved the way for calculus. He became one of the world's leading mathematicians, but published little of his work, mainly because he was not willing to spend the time required to bring it into publishable form.

His most long-lasting influence was in number theory, where he challenged other mathematicians to prove a series of theorems and solve various problems. Among these was the (misnamed) 'Pell equation' $nx^2 + 1 = y^2$, and the statement that the sum of two non-zero perfect cubes cannot be a perfect cube. This is a special case of a more general conjecture, 'Fermat's Last Theorem', in which cubes are replaced by nth powers for any $n \geq 3$.

He died in 1665, just two days after concluding a legal case.

and currently only, proof was derived by Andrew Wiles in 1994; it uses advanced abstract methods that did not exist until the late 20th century.

After Fermat, several major mathematicians worked in number theory, notably Euler and Lagrange. Most of the theorems that Fermat had stated but not proved were polished off during this period.

Gauss

The next big advance in number theory was made by Gauss, who published his masterpiece, the *Disquisitiones Arithmeticae (Investigations in Arithmetic)* in 1801. This book propelled the theory of numbers to the centre of the mathematical stage. From then on, number theory was a core component of the mathematical mainstream. Gauss mainly focused on his own, new, work, but he also set up the foundations of number theory and systematized the ideas of his predecessors.

The most important of these foundational changes was a very simple but powerful idea: *modular arithmetic*. Gauss discovered a new type of number system, analogous to integers but differing in one key respect: some particular number, known as the *modulus*, was identified with the number zero. This curious idea turned out to be fundamental to our understanding of divisibility properties of ordinary integers.

Here is Gauss's idea. Given an integer m, say that a and b are *congruent to the modulus* m, denoted

$$a \equiv b (\text{mod } m)$$

if the difference $a - b$ is exactly divisible by m. Then arithmetic to the modulus m works exactly the same as ordinary arithmetic, except that we may replace m by 0 anywhere in the calculation. So, any multiple of m can be ignored.

The phrase 'clock arithmetic' is often used to capture the spirit of Gauss's idea. On a clock, the number 12 is effectively the same as 0 because the hours repeat after 12 steps (24 in continental Europe and military activities). Seven hours after

Carl Friedrich Gauss

1777–1855

Gauss was highly precocious, allegedly correcting his father's arithmetic when aged three. In 1792, with financial assistance from the Duke of Brunswick-Wolfenbüttel, Gauss went to Brunswick's Collegium Carolinum. There he made several major mathematical discoveries, including the law of quadratic reciprocity and the prime number theorem, but did not prove them. From 1795–98 he studied at Göttingen, where he discovered how to construct a regular 17-sided polygon with ruler and compass. His *Disquisitiones Arithmeticae*, the most important work in number theory to date, was published in 1801.

Gauss's public reputation, however, rested on an astronomical prediction. In 1801 Giuseppe Piazzi discovered the first asteroid: Ceres. The observations were so sparse that astronomers were worried that they might not find it again when it reappeared from behind the Sun. Several astronomers predicted where it would reappear; so did Gauss. Only Gauss was right. In fact, Gauss had used a method of his own invention, now called the 'method of least squares', to derive accurate results from limited observations. He did not reveal this technique at the time, but it has since become fundamental in statistics and observational science.

In 1805 Gauss married Johanna Ostoff, whom he loved dearly, and in 1807 he left Brunswick to become director of the Göttingen observatory. In 1808 his father died, and Johanna died in 1809 after giving birth to their second son. Soon after, the son died too.

Despite these personal tragedies, Gauss continued his research, and in 1809 he published his *Theoria Motus Corporum Coelestium in Sectionibus Conicis Solem Ambientium*, a major contribution to celestial mechanics. He married again, to Minna, a close friend of Johanna's, but the marriage was more out of convenience than love.

Around 1816 Gauss wrote a review of deductions of the parallel axiom from the other axioms of Euclid, in which he hinted at an opinion he had probably held since 1800, the possibility of a logically consistent geometry that differed from Euclid's.

In 1818 he was placed in charge of a geodetic survey of Hanover, making serious contributions to the methods employed in surveying. In 1831, after the death of Minna, Gauss began working with the physicist Wilhelm Weber on the Earth's magnetic field.

They discovered what are now called Kirchhoff's laws for electrical circuits, and constructed a crude but effective telegraph. When Weber was forced to leave Göttingen in 1837, Gauss's scientific work went into decline, though he remained interested in the work of others, notably Ferdinand Eisenstein and Georg Bernhard Riemann. He died peacefully in his sleep.

6 o'clock is not 13 o'clock, but 1 o'clock, and in Gauss's system $13 \equiv 1 \pmod{12}$. So modular arithmetic is like a clock that takes m hours to go full circle. Not surprisingly, modular arithmetic crops up whenever mathematicians look at things that change in repetitive cycles.

The *Disquisitiones Arithmeticae* used modular arithmetic as the basis for deeper ideas, and we mention three.

The bulk of the book is a far-reaching extension of Fermat's observations that primes of the form 4k + 1 are a sum of two squares, whereas those of the form 4k – 1 are not. Gauss restated this result as a characterization of integers that can be written in the form $x^2 + y^2$, with x and y integers. Then he asked what happens if instead of this formula we use a general *quadratic form*, $ax^2 + bxy + cy^2$. His theorems are too technical to discuss, but he obtained an almost complete understanding of this question.

Another topic is the law of quadratic reciprocity, which intrigued and perplexed Gauss for many years. The starting point is a simple question: what do perfect squares look like to a given modulus?

For instance, suppose that the modulus is 11. Then the possible perfect squares (of the numbers less than 11) are

$$0 \quad 1 \quad 4 \quad 9 \quad 16 \quad 25 \quad 36 \quad 49 \quad 64 \quad 81 \quad 100$$

which, when reduced (mod 11), yield

$$0 \quad 1 \quad 3 \quad 4 \quad 5 \quad 9$$

with each non-zero number appearing twice. These numbers are the *quadratic residues*, mod 11.

The key to this question is to look at prime numbers. If p and q are primes, when is q a square (mod p)? Gauss discovered that while there is no simple way to answer that question directly, it bears a remarkable relation to another question: when is p a square (mod q)? For example, the list of quadratic residues above shows that $q = 5$ is a square modulo $p = 11$. It is also true that 11 is a square modulo 5 – because $11 \equiv 1 \pmod 5$ and $1 = 1^2$. So here both questions have the same answer.

Gauss proved that this law of reciprocity holds for any pair of odd primes, except when both primes are of the form $4k - 1$, in which case the two questions always have opposite answers. That is: for any odd primes p and q,

q is a square (mod p) if and only if p is a square (mod q),

unless both p and q are of the form $4k - 1$, in which case

q is a square (mod p) if and only if p is *not* a square (mod q).

Initially Gauss was unaware that this was not a new observation: Euler had noticed the same pattern. But unlike Euler, Gauss managed to prove that it is always true. The proof was very difficult, and it took Gauss several years to fill one small but crucial gap.

A third topic in the *Disquisitiones* is the discovery that had convinced Gauss to become a mathematician at the age of 19: a geometric construction for the regular 17-gon (a polygon with 17 sides). Euclid provided constructions, using unmarked ruler and compass, for regular polygons with three, five and 15 sides; he also knew that those numbers could be repeatedly doubled by bisecting angles, yielding regular polygons with four, six, eight and 10 sides, and so on. But Euclid gave no constructions for 7-sided polygons, 9-sided ones, or indeed any other numbers than the ones just listed. For some two thousand years, the mathematical world assumed that Euclid had said the last word, and no other regular polygons were constructable. Gauss proved them wrong.

It is easy to see that the main problem is constructing regular p-gons when p is prime. Gauss pointed out that such a construction is equivalent to solving the algebraic equation

$$x^{p-1} + x^{p-2} + x^{p-3} + \cdots + x^2 + x + 1 = 0$$

Now, a ruler-and-compass construction can be viewed, thanks to coordinate geometry, as a sequence of quadratic equations. If a construction of this kind exists, it follows (not entirely trivially) that $p - 1$ must be a power of 2.

The Greek cases $p = 3$ and 5 satisfy this condition: here $p - 1 =$ 2 and 4, respectively. But they are not the only such primes. For instance $17 - 1 = 16$ is a power of 2. This does not yet prove that the 17-gon is constructable, but it provides a strong hint, and Gauss managed to find an explicit reduction of his 16th degree equation to a series of quadratics. He stated, but did not prove, that a construction is possible whenever $p - 1$ is a power of 2 (still requiring p to be *prime*), and it is impossible for all other primes. The proof was soon completed by others.

These special primes are called *Fermat primes*, because they were studied by Fermat. He observed that if p is a prime and $p - 1 = 2^k$, then k must itself be a power of 2. He noted the first few Fermat primes: 2, 3, 5, 17, 257 and 65,537. He conjectured that numbers of the form $2^{2^m} + 1$ are always prime, but this was wrong. Euler

What number theory did for them

One of the earliest practical applications of number theory occurs in gears. If two cogs are placed together so that their teeth mesh, and one cog has *m* teeth, the other *n* teeth, then the movement of the cogs is related to these numbers. For instance, suppose one cog has 30 teeth and the other has 7. If I turn the big cog exactly once, what does the smaller one do? It returns to the initial position after 7, 14, 21 and 28 steps. So, the final 2 steps, to make 30, advance it by just 2 steps. This number turns up because it is the remainder on dividing 30 by 7. So the motion of cogs is a mechanical representation of division with remainder, and this is the basis of modular arithmetic.

Cogwheels were used by ancient Greek craftsmen to design a remarkable device, the Antikythera mechanism. In 1900 the sponge diver Elias Stadiatis found a formless lump of corroded rock in a wreck of 65 BC near the island of Antikythera, some 40 metres (140 feet) down. In 1902 the archaeologist Valerios Stais noticed that the rock contained a gear wheel, and was actually the remains of a complicated bronze mechanism. It was inscribed with words in the Greek alphabet.

The mechanism's function has been worked out from its structure and its inscriptions, and it turns out to be an astronomical calculator. There are more than 30 gear wheels – the latest reconstruction, in 2006, suggests there were originally 37. The numbers of cogs correspond to important astronomical ratios. In particular, two cogs have 53 teeth – a difficult number to manufacture – and this number comes from the rate at which the Moon's furthest point from the Earth rotates. All the prime factors of the numbers of teeth are based on two classical astronomical cycles, the Metonic and Saros cycles. X-ray analysis has revealed new inscriptions and made them readable, and it is now certain that the device was used to predict the positions of the Sun, the Moon and probably the then-known planets. The inscriptions date to around 150–100 BC.

The Antikythera mechanism is of a sophisticated design, and appears to incorporate Hipparchus's theory of the motion of the Moon. It may well have been built by one of his students, or at least with their aid. It was probably an executive toy for a royal personage, rather than a practical instrument, which may explain its exquisite design and manufacture.

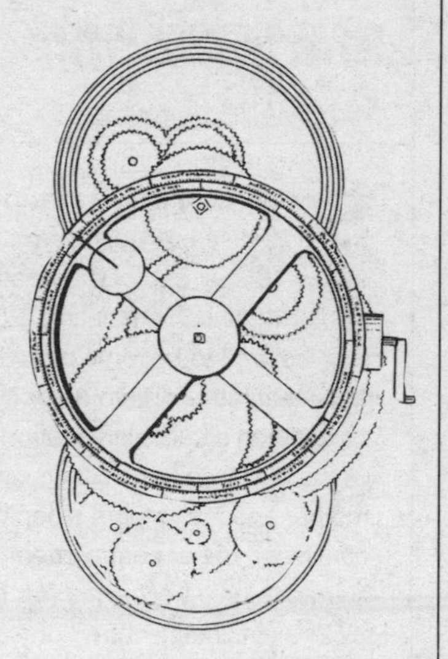

A reconstruction of the Antikythera mechanism

discovered that when m = 5 there is a factor 641.

It follows that there must also exist ruler-and-compass constructions for the regular 257-gon and 65,537-gon. F.J. Richelot constructed the regular 257-gon in 1832, and his work is correct. J. Hermes spent ten years working on the 65,537-gon, and completed his construction in 1894. Recent studies suggest there are mistakes.

Number theory started to become mathematically interesting with the work of Fermat, who spotted many of the important patterns concealed in the strange and puzzling behaviour of whole numbers. His annoying tendency not to supply proofs was put right by Euler, Lagrange and a few less prominent figures, with the sole exception of his Last Theorem, but number theory seemed to consist of isolated theorems – often deep and difficult, but not very closely connected to each other.

Marie-Sophie Germain

1776–1831

Sophie Germain was the daughter of the silk merchant Ambroise-François Germain and Marie-Madelaine Gruguelin. At age 13 she read of the death of Archimedes, killed by a Roman soldier while contemplating a geometric diagram in the sand, and was inspired to become a mathematician. Despite her parents' well-meaning efforts to deter her – mathematics was not then considered a suitable vocation for a young lady – she read the works of Newton and Euler, wrapped in a blanket while her mother and father were sleeping. When her parents became convinced of her commitment to mathematics, they relented and started to help her, giving her financial support throughout her life.

She obtained lecture notes from the École Polytechnique, and wrote to Lagrange with some original work of her own under the name 'Monsieur LeBlanc'. Lagrange, impressed, eventually discovered that the writer was a woman, and had the good sense to encourage her and become her sponsor. The two worked together, and some of her results were included in a later edition of Legendre's 1798 *Essai sur le Théorie des Nombres*.

Her most celebrated correspondent was Gauss. Sophie studied the *Disquisitiones Arithmeticae*, and from 1804 to 1809 she wrote a number of letters to its author, again concealing her gender by using the name LeBlanc. Gauss praised LeBlanc's work in letters to other mathematicians. In 1806 he discovered that LeBlanc was actually female, when the French occupied Braunschweig. Concerned that Gauss might suffer the same fate as Archimedes, Sophie contacted a family friend who was a senior officer in the French Army, General Pernety. Gauss learned of this, and discovered that LeBlanc was actually Sophie.

Sophie need not have worried. Gauss was even more impressed, and wrote to her: 'But how to describe to you my admiration and astonishment at seeing my esteemed correspondent Monsieur Le Blanc metamorphose himself into this illustrious personage ... When a person of the sex which, according to our customs and prejudices, must encounter infinitely more difficulties than men to familiarize herself with these thorny researches, succeeds nevertheless in surmounting these obstacles and penetrating the most obscure parts of them, then without doubt she must have the noblest courage, quite extraordinary talents and superior genius.'

Sophie obtained some results on Fermat's Last Theorem, the best available until 1840. Between 1810 and 1820 she worked on the vibrations of surfaces, a problem posed by the Institut de France. In particular, an explanation was sought for 'Chladni patterns' – symmetric patterns that appear if sand is sprinkled on a metal plate, which is then caused to vibrate using a violin bow. On her third attempt she was awarded a gold medal, although for unknown reasons, possibly a protest about the unfair treatment of women scientists, she did not appear at the award ceremony.

In 1829 she developed breast cancer, but continued to work on number theory and the curvature of surfaces until her death two years later.

All that changed when Gauss got in on the act and devised general conceptual foundations for number theory, such as modular arithmetic. He also related number theory to geometry with his work on regular polygons. From that moment, number theory became a major strand in the tapestry of mathematics.

Gauss's insights led to the recognition of new kinds of structure in mathematics – new number systems, such as the integers mod n, and new operations, such as the composition of quadratic forms. With hindsight, the number theory of the late 18th and early 19th centuries led to the abstract algebra of the late 19th and 20th

What number theory does for us

Number theory forms the basis of many important security codes used for Internet commerce. The best known such code is the RSA (Ronald Rivest, Adi Shamir and Leonard Adleman) cryptosystem, which has the surprising feature that the method for putting messages into code can be made public without giving away the reverse procedure of decoding the message.

Suppose Alice wants to send a secret message to Bob. Before doing this, they agree on two large primes p and q (having at least a hundred digits) and multiply them together to get $M = pq$. They can make this number public if they wish. They also compute $K = (p - 1)(q - 1)$, but keep this secret.

Now Alice represents her message as a number x in the range 0 to M (or a series of such numbers if it's a long message). To encode the message she chooses some number a, which has no factors in common with K, and computes $y \equiv x^a \pmod{M}$. The number a must be known to Bob, and can also be made public.

To decode messages, Bob has to know a number b such that $ab \equiv 1 \bmod K$. This number (which exists and is unique) is kept secret. To decode y, Bob computes

$$y^b \pmod{M}.$$

Why does this decode? Because

$$y^b \equiv (x^a)^b \equiv x^{ab} \equiv x^1 \equiv x \pmod{M},$$

using a generalization of Fermat's Little Theorem due to Euler.

This method is practical because there are efficient tests to find large primes. However, there are no known methods for finding the prime factors of a large number efficiently. So telling people the product pq does not help them find p and q, and without those, they cannot work out the value of b, needed to decode the message.

centuries. Mathematicians were starting to enlarge the range of concepts and structures that were acceptable objects of study. Despite its specialized subject matter, the *Disquisitiones Arithmeticae* marks a significant milestone in the development of the modern approach to the whole of mathematics. This is one of the reasons why Gauss is rated so highly by mathematicians.

Until the late 20th century, number theory remained a branch of pure mathematics – interesting in its own right, and because of its numerous applications within mathematics itself, but of little real significance to the outside world. All that changed with the invention of digital communications in the late 20th century. Since communication then depended on numbers, it is hardly a surprise that number theory came to the fore in those areas of application. It often takes time for a good mathematical idea to acquire practical importance – sometimes hundreds of years – but eventually most topics that mathematicians find significant for their own sake turn out to be valuable in the real world too.

CHAPTER 8

The System of the World

The invention of calculus

The most significant single advance in the history of mathematics was calculus, invented independently around 1680 by Isaac Newton and Gottfried Leibniz. Leibniz published first, but Newton – egged on by over-patriotic friends – claimed priority and portrayed Leibniz as a plagiarist. The row soured relations between English mathematicians and those of continental Europe for a century, and the English were the main losers.

The system of the world

Even though Leibniz probably deserves priority, Newton turned calculus into a central technique of the budding subject of mathematical physics, humanity's most effective known route to the understanding of the natural world. Newton called his theory 'The System of the World'. This may not have been terribly modest, but it was a pretty fair description. Before Newton, human understanding of patterns in nature consisted mainly of the ideas of Galileo about moving bodies, in particular the parabolic trajectory of an object such as a cannonball, and Kepler's discovery that Mars follows an ellipse through the heavens. After Newton, mathematical patterns governed almost everything in the physical world: the movement of terrestrial and heavenly bodies, the flow of air and water, the transmission of heat, light and sound, and the force of gravity.

Curiously, though, Newton's main publication on the mathematical laws of nature, his *Principia Mathematica*, does not mention calculus at all; instead, it relies on the clever application of geometry in the style of the ancient Greeks. But appearances are deceptive: unpublished documents known as the *Portsmouth Papers* show that when he was working on the *Principia*, Newton already had the main ideas of calculus. It is likely that Newton *used* the methods of calculus to make many of his discoveries, but chose not to present them that way. His version of calculus was published after his death in the *Method of Fluxions* of 1732.

Calculus

What is calculus? The methods of Newton and Leibniz are more easily understood if we preview the main ideas. Calculus is the mathematics of instantaneous rates of change – how rapidly is some particular quantity changing *at this very instant*? For a physical example: a train is moving along a track: how fast is it going *right now*? Calculus has two main branches. *Differential calculus* provides methods for calculating rates of change, and it has many geometric applications, in particular finding tangents to curves. *Integral calculus* does the opposite: given the rate of change of some quantity, it specifies the quantity itself. Geometric applications of integral calculus include the computation of areas and volumes. Perhaps the most significant discovery is this unexpected connection between two apparently unrelated classical geometric questions: finding tangents to a curve and finding areas.

Calculus is about *functions*: procedures that take some general number and calculate an associated number. The procedure is usually specified by a formula, assigning to a given number x (possibly in some specific range) an associated number $f(x)$. Examples include the square root function $f(x) = \sqrt{x}$ (which requires x to be positive) and the square function $f(x) = x^2$ (where there is no restriction on x).

The first key idea of calculus is *differentiation*, which obtains the *derivative* of a function. The derivative is the rate at which $f(x)$ is changing, compared to how x is changing – the *rate of change* of $f(x)$ with respect to x.

Geometrically, the rate of change is the slope of the tangent to the graph of f at the value x. It can be approximated by finding the slope of the *secant* – a line that cuts the graph of f at two nearby points, corresponding to x and $x + h$, respectively, where h is small. The slope of the secant is

$$\frac{f(x + h) - f(x)}{h}$$

Now suppose that h becomes very small. Then the secant approaches the tangent to the graph at x. So in some sense the required slope – the derivative of f at x – is the limit of this expression as h becomes arbitrarily small.

Let's try this calculation with a simple example, $f(x) = x^2$. Now

$$\frac{f(x + h) - f(x)}{h} = \frac{(x + h)^2 - x^2}{h} = \frac{x^2 + 2hx + h^2 - x^2}{h} = 2x + h$$

As h becomes very, very small, the slope $2x + h$ gets closer and closer to $2x$. So the derivative of f is the function g for which $g(x) = 2x$. The main conceptual issue here is to define what we mean by limit. It took more than a century to find a logical definition.

The other key idea in calculus is that of *integration*. This is most easily viewed as the reverse process to differentiation. Thus the integral of g, written

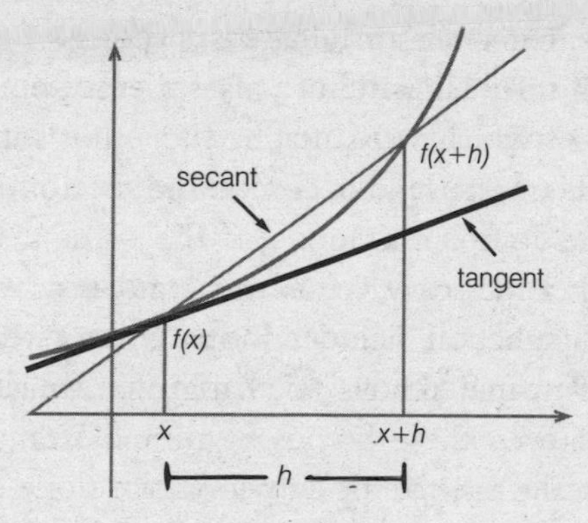

Geometry of approximations to the derivative

$$\int g(x)dx$$

is whichever function $f(x)$ has derivative $g(x)$. For instance, because the derivative of $f(x) = x^2$ is $g(x) = 2x$, the integral of $g(x) = 2x$ is $f(x) = x^2$. In symbols,

$$\int 2x\,dx = x^2$$

The need for Calculus

Inspiration for the invention of calculus came from two directions. Within pure mathematics, differential calculus evolved from methods for finding tangents to curves, and integral calculus evolved from methods for calculating the areas of plane shapes and the volumes of solids. But the main stimulus towards calculus came from physics – the growing realization that nature has patterns. For reasons we still do not really understand, many of the fundamental patterns in nature involve rates of change. So they make sense, and can be discovered, only through calculus.

Prior to the Renaissance, the most accurate model of the motion of the Sun, Moon and planets was that of Ptolemy. In his model, the Earth was fixed, and everything else – in particular, the Sun – revolved around it on a series of (real or imaginary, depending on taste) circles. The circles originated as spheres in the work of the Greek astronomer Hipparchus; his spheres spun about gigantic axles, some of which were attached to other spheres and moved along with them. This kind of compound motion seemed necessary to model the complex motions of the planets. Recall that some planets, such as Mercury, Venus and Mars, seemed to travel along complicated paths that included loops. Others – Jupiter and Saturn were the only other planets known at that time – behaved more sedately, but even these bodies exhibited strange irregularities, known since the time of the Babylonians.

We have already met Ptolemy's system, known as *epicycles*, which replaced the spheres by circles, but retained the compound motion. Hipparchus's model was not terribly accurate, compared to observations, but Ptolemy's model fitted the observations very accurately indeed, and for over a thousand years it was seen as the last word on the topic. His writings, translated into Arabic as the *Almagest*, were used by astronomers in many cultures.

God v science

Even the *Almagest*, however, failed to agree with all planetary movements. Moreover, it was rather complicated. Around the year 1000, a few Arabian and European thinkers began to wonder whether the daily motion of the Sun might be explained by a rotating Earth, and some of them also played with the idea that the Earth revolves round the Sun. But little came of these speculations at the time.

In Renaissance Europe, however, the scientific attitude began to take root, and one of the first casualties was religious dogma. At that time, the Roman Catholic Church exerted substantial control over its adherents' view of the universe. It wasn't just that the existence of the universe, and its daily unfolding, were credited to the Christian God. The point was that the nature of the universe was believed to correspond to a very literal reading of the Bible.

The Earth was therefore seen as the centre of all things, the solid ground around which the heavens revolved. And human beings were the pinnacle of creation, the reason for the universe's existence.

No scientific observation can ever disprove the existence of some invisible, unknowable creator. But observations can – and did – debunk the view of the Earth as the centre of the universe. And this caused a huge fuss, and got a lot of innocent people killed, sometimes in hideously cruel ways.

Copernicus

The fat hit the fire in 1543, when the Polish scholar Nicholas Copernicus published an astonishing, original and somewhat heretical book: *On the Revolutions of the Heavenly Spheres*. Like Ptolemy, he used epicycles for accuracy. Unlike Ptolemy, he placed the Sun at the centre, while everything else, including the Earth, but excluding the Moon, turned around the Sun. The Moon alone revolved around the Earth.

Copernicus's main reason for this radical proposal was pragmatic: it replaced Ptolemy's 77 epicycles by a mere 34. Among the epicycles envisaged by Ptolemy, there were many repetitions of a particular circle: circles with that specific size, and speed of rotation, kept appearing, associated with many distinct celestial bodies. Copernicus realized that if all these epicycles were transferred to the Earth, only one of them would be needed. We now interpret this in terms of the motion of the planets relative to the Earth. If we mistakenly assume the Earth is fixed, as it seems to be to a naive observer, then the motion of the Earth round the Sun becomes transferred to all of the planets as an additional epicycle.

Another advantage of Copernicus's theory was that it treated all the planets in exactly the same manner. Ptolemy needed different mechanisms to explain the inner planets and the outer ones. Now, the only difference was that the inner planets were closer to the Sun than the Earth was, while the outer planets were further away. It all made excellent sense – but on the whole it was rejected, for a variety of reasons, not all of them religious.

Copernicus's theory was complicated, unfamiliar and his book was difficult to read. Tycho Brahe, one of the best astronomical observers of the period, found discrepancies between Copernicus's heliocentric theory and some subtle observations, which also disagreed with Ptolemy's theory; he tried to find a better compromise.

Kepler

When Brahe died, his papers were inherited by Kepler, who spent years analysing the observations, looking for patterns. Kepler was something of a mystic, in the Pythagorean tradition, and he tended to impose artificial patterns on observational data. The most famous of these abortive attempts to find regularities in the heavens was his beautiful but utterly misguided explanation of the spacing of the planets in terms of the regular solids. In his day, the known planets were six in number: Mercury, Venus, Earth, Mars, Jupiter and Saturn. Kepler wondered whether their distances from the Sun had a geometric pattern. Moreover, he wondered why there were six planets. He noticed that six planets leave room for five intervening shapes, and since there were exactly five regular solids, this would explain the limit of six planets. He came up with a series of six spheres lying one inside the other, each one bearing the orbit of a planet around its equator. Between the spheres, nestling tightly outside one sphere and inside the next, he placed the five solids, in the order

Mercury
Octahedron
Venus
Icosahedron

Earth

Dodecahedron

Mars

Tetrahedron

Jupiter

Cube

Saturn

The numbers fitted reasonably well, especially given the limited accuracy of observations at that time. But there are 120 different ways to rearrange the five solids, which between them give an awful lot of different spacings. It is hardly surprising that one of these was in reasonably close agreement with reality. The later discovery of more planets knocked this particular piece of pattern-seeking firmly on the head, consigning it to the waste bin of history.

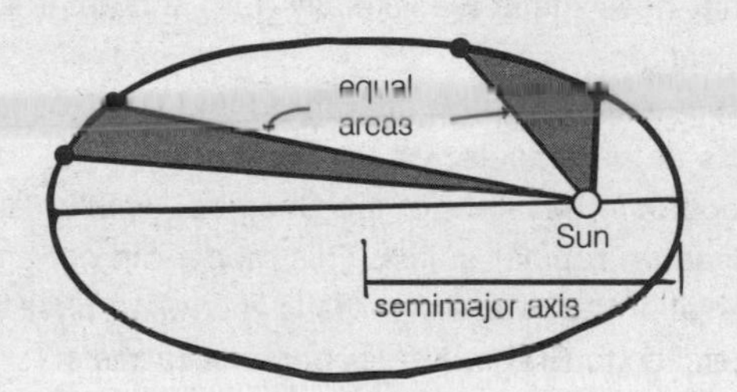

Planet moves over given time interval

Along the way, though, Kepler discovered some patterns that we still recognize as genuine, now called *Kepler's Laws of Planetary Motion*. He extracted them, over some twenty years of calculation, from Brahe's observations of Mars. The laws state:

(i) Planets move round the Sun in elliptical orbits.
(ii) Planets sweep out equal areas in equal times.
(iii) The square of the period of revolution of any planet is proportional to the cube of its average distance from the Sun.

Johannes Kepler
1571–1630

Kepler was the son of a mercenary and an innkeeper's daughter. As a child, he lived with his mother in his grandfather's inn after the death of his father, probably in a war between the Netherlands and the Holy Roman Empire. He was mathematically precocious, and in 1589 he studied astronomy under Michael Maestlin at the University of Tübingen. Here he came to grips with the Ptolemaic system. Most astronomers of the period were more concerned with calculating orbits than in asking how the planets really moved, but from the beginning Kepler was interested in the precise paths followed by the planets, rather than the proposed system of epicycles. He became acquainted with the Copernican system, and was quickly convinced that it was literally true, and not just a mathematical trick.

In 1596 he made his first attempt to find patterns in the movements of the planets, via his *Mysterium Cosmographicum* (Mystery of the Cosmos), with its strange model based on regular solids. This model did not agree well with observations, so Kepler wrote to a leading observational astronomer, Tycho Brahe. Kepler became Brahe's mathematical assistant, and was set to work on the orbit of Mars. After Brahe's death, he continued to work on the problem. Brahe had left a wealth of data, and Kepler struggled to fit a sensible orbit to them. The surviving calculations occupy nearly 1000 pages, which Kepler referred to as 'my war with Mars'. His final orbit was so precise that the only difference from modern data arises from minute drifting of the orbit over the intervening centuries.

1611 was a bad year. Kepler's son died aged seven. Next, his wife died. Then Emperor Rudolf, who tolerated Protestants, abdicated, and Kepler was forced to leave Prague. In 1613

Kepler remarried, and a problem that occurred to him during their wedding celebrations led to the writing of his *New Stereometry of Wine Barrels* of 1615.

In 1619 he published *Harmonices Mundi* (The Harmony of the World), a sequel to his Mystery of the Cosmos. The book contained a wealth of new mathematics, including tiling patterns and polyhedra. It also formulated the third law of planetary motion. While he was writing the book, his mother was accused of being a witch. With the aid of the law faculty at Tübingen, she was eventually freed, partly because the prosecutors had not followed the correct legal procedures for torture.

The most unorthodox feature of Kepler's work is that he discarded the classical circle (allegedly the most perfect shape possible) in favour of the ellipse. He did so with some reluctance, saying himself that he only settled on the ellipse when everything else had been eliminated. There is no particular reason to expect these three laws to bear any closer relation to reality than the hypothetical arrangement of regular solids, but as it happened, the three laws were of real scientific significance.

Galileo

Another major figure of the period was Galileo Galilei, who discovered mathematical regularities in the movement of a pendulum and in falling bodies. In 1589, as professor of mathematics at the University of Pisa, he carried out experiments on bodies rolling down an inclined plane, but did not publish his results. It was at this time that he realized the importance of controlled experiments in the study of natural phenomena, an idea that is now fundamental to all science. He took up astronomy, making a series of fundamental discoveries, which eventually led him to espouse the Copernican theory of the Sun as the centre of

Galileo Galilei

1564–1642

Galileo was the son of Vincenzo Galilei, a music teacher who had performed experiments with strings to validate his musical theories. At the age of ten Galileo went to a monastery at Vallombrosa to be educated, with a view to becoming a doctor. But Galileo wasn't really interested in medicine, and spent his time doing mathematics and natural philosophy – what we now call science.

In 1589 Galileo became professor of mathematics at the University of Pisa. In 1591 he took up a better-paid position in Padua, where he taught Euclidean geometry and astronomy to medical students. At that time doctors made use of astrology in their treatment of patients, so these topics were a necessary part of the curriculum.

Learning about the invention of the telescope, Galileo built one for himself and became so proficient that he gave his methods to the Venetian Senate, granting them sole rights in their use in return for a salary increase. In 1609 Galileo observed the heavens, making discovery after discovery: four of Jupiter's moons, individual stars within the Milky Way, mountains on the Moon. He presented a telescope to Cosimo de Medici, Grand Duke of Tuscany, and soon became the Duke's chief mathematician.

He discovered the existence of sunspots and published this observation in 1612. By now his astronomical discoveries had convinced him of the truth of Copernicus's heliocentric theory, and in 1616 he made his views explicit in a letter to the Grand Duchess Christina, saying that the Copernican theory represents physical reality and is not just a convenient way to simplify calculations.

At this time, Pope Paul V ordered the Inquisition to decide on the truth or falsity of the heliocentric theory, and they declared it false. Galileo was instructed not to advocate the theory, but a new pope was elected, Urban VIII, who seemed more relaxed about the issue, so Galileo did not take the prohibition very seriously. In 1623 he published *Il Saggiatore* (*The Assayer*), dedicating it to Urban. In it, he made a famous statement that the universe 'is written in the language of mathematics, and its characters are triangles, circles and other geometric figures, without which it is humanly impossible to understand a single word of it'.

In 1630 Galileo asked permission to publish another book, *Dialogue Concerning the Two Chief Systems of the World*, about the geocentric and heliocentric theories. In 1632, when permission arrived from Florence (but not Rome), he went ahead. The book claimed to prove that the Earth moves, the main evidence being the tides. In fact Galileo's theory of tides was completely wrong, but the Church authorities saw the book as theological dynamite and the Inquisition banned it, summoning Galileo to Rome to be tried for heresy. He was found guilty, but escaped with a sentence of life imprisonment, in the form of house arrest. In this he fared better than many other heretics, for whom being burnt at the stake was a common punishment. While under house arrest he wrote the *Discourses*, explaining his work on moving bodies to the outside world. It was smuggled out of Italy and published in Holland.

the solar system. This set him on a collision course with the Church, and he was eventually tried for heresy and placed under house arrest.

During the last few years of his life, his health failing, he wrote *Discourses and Mathematical Demonstrations Concerning the Two New Sciences*, explaining his work on the motion of bodies on inclined planes. He stated that the distance an initially stationary body moves under uniform acceleration is proportional to the square of the time. This law is the basis of his earlier discovery that a projectile follows a parabolic path. Together with Kepler's laws of planetary motion, it brought into being a new subject: *mechanics*, the mathematical study of moving bodies.

That's the physical astronomical background that led up to calculus. Next, we'll look at the mathematical background.

The invention of calculus

The invention of calculus was the outcome of a series of earlier investigations of what seem to be unrelated problems, but which possess a hidden unity. These included calculating the instantaneous velocity of a moving object from the distance it has travelled at any given time, finding the tangent to a curve, finding the length of a curve, finding the maximum and minimum values of a variable quantity, finding the area of some shape in the plane and the volume of some solid in space. Some important ideas and examples were developed by Fermat, Descartes and the more obscure Englishman, Isaac Barrow, but the methods remained special to particular problems. A general method was needed.

Leibniz

The first real breakthrough was made by Gottfried Wilhelm Leibniz, a lawyer by profession, who devoted much of his life to mathematics, logic, philosophy, history and many branches of science. Around 1673 he began work on the classical problem of finding the tangent to a curve, and noticed that this was in effect the inverse problem to that of finding areas and volumes. The latter boiled down to finding a curve given its tangents; the former problem was exactly the reverse.

Leibniz used this connection to define what, in effect, were integrals, using the abbreviation, *omn* (an abbreviation of *omnia*, the Latin word for 'all'). Thus we find, in his manuscript papers, formulas such as

$$\text{omn } x^2 = \frac{x^3}{3}$$

By 1675 he had replaced *omn* by the symbol $\int$ still used today, which is an old-fashioned elongated letter s, standing for sum. He worked in terms of small increments dx and dy to the quantities x and y, and used their ratio dy/dx to determine the rate of change of y as a function of x. Essentially, if f is a function then Leibniz wrote

$$dy = f(x + dx) - f(x)$$

so that

$$\frac{dy}{dx} = \frac{f(x + dx) - f(x)}{dx}$$

which is the usual secant approximation to the slope of the tangent.

Leibniz recognized that this notation has its problems. If dy and dx are non-zero then dy/dx is not the instantaneous rate of change of y, but an approximation. He tried to circumvent this problem by assuming dx and dy to be infinitesimally small. An *infinitesimal* is a non-zero number that is smaller than any other non-zero number. Unfortunately, it is easy to see that no such number can exist (half an infinitesimal is also non-zero, but smaller) so this approach does little more than displace the problem elsewhere.

By 1676 Leibniz knew how to integrate and differentiate any power of x, writing the formula

$$dx^n = nx^{n-1}dx$$

which we would now write as

$$\frac{d}{dx}x^n = nx^{n-1}$$

In 1677 he derived rules for differentiating the sum, product and quotient of two functions, and by 1680 he had obtained the formula for the length of an arc of a curve, and the volume of a solid of revolution, as integrals of various related quantities.

Although we know these facts, and the associated dates, from his unpublished notes, he first published his ideas on calculus rather later, in 1684. Jakob and Johann Bernoulli found this paper rather obscure, describing it as 'an enigma rather than an explanation'. With hindsight, we see that by that time Leibniz had discovered a significant part of basic calculus, with applications to complicated curves like the cycloid, and a sound grasp of concepts such as curvature. Unfortunately, his writings were fragmented and virtually unreadable.

Newton

The other creator of calculus was Isaac Newton. Two of his friends, Isaac Barrow and Edmond Halley, came to recognize his remarkable abilities, and encouraged him to publish his work. Newton disliked being criticized, and when in 1672 he published his ideas about light, his work provoked a storm of criticism, which reinforced his reluctance to commit his thoughts to print. Nevertheless, he continued to publish sporadically, and wrote two books. In private he continued developing his ideas about gravity, and in 1684 Halley tried to convince Newton to publish the work. But aside from Newton's general misgivings about criticism, there was a technical obstacle. He had been forced to model planets as point particles, with non-zero mass but zero size, which he felt was unrealistic and would invite criticism. He wanted to replace these unrealistic points by solid spheres, but he could not prove that the gravitational attraction of a sphere is the same as that of a point particle of the same mass.

In 1686 he succeeded in filling the gap, and the *Principia* saw the light of day in 1687. It contained many novel ideas. The most important were mathematical laws of motion, extending the work of Galileo, and gravity, based on the laws found by Kepler.

Newton's main law of motion (there are some subsidiary ones) states that the acceleration of a moving body, multiplied by its mass, is equal to the force that acts on the body. Now velocity is the derivative of position, and acceleration is the derivative of velocity.

Isaac Newton
1642–1727

Newton lived on a farm in the tiny village of Woolsthorpe, in Lincolnshire. His father had died two months before he was born, and his mother managed the farm. He was educated in very ordinary local schools, and exhibited no special talent of any kind, except for a facility with mechanical toys. He once made a hot air balloon and tested it out with the family cat as pilot; the balloon and the cat were never seen again. He went to Trinity College at Cambridge University, having done reasonably well in most of his examinations – except geometry. As an undergraduate, he made no great impact.

The Plague

Then, in 1665, the great plague began to devastate London and the surrounding area, and the students were sent home before the same thing happened in Cambridge. Back in the family farmhouse, Newton began thinking much more deeply about scientific and mathematical matters.

Gravity

During 1665–66 he devised his law of gravity to explain planetary motion, developed the laws of mechanics to explain and analyse any kind of moving body or particle, invented both differential and integral calculus, and made major advances in optics. Characteristically, he published none of this work, returning quietly to Trinity to take his master's degree and being elected a fellow of the college. Then he secured the position of Lucasian Professor of Mathematics, when the incumbent, Barrow, resigned in 1669. He gave very ordinary lectures, rather badly, and very few undergraduates went to them.

So even to *state* Newton's Law we need the *second derivative* (the derivative of the derivative) of position with respect to time, nowadays written

$$\frac{d^2x}{dt^2}$$

Newton wrote two dots over the top of *x* instead ($\ddot{x}$).

The law of gravity states that any two particles of matter attract each other with a force that is proportional to their masses, and inversely proportional to the square of the distance between them. So, for example, the force attracting the Earth to the Moon would become one quarter as great if the Moon were removed to twice its distance, or one ninth as great if its distance were tripled. Again, because this law is about forces, it involves the second derivative of position.

Newton deduced this law from Kepler's three laws of planetary motion. The published deduction was a masterpiece of classical Euclidean geometry. Newton chose this style of presentation because it involved familiar mathematics, and so could not easily be criticized. But many aspects of the *Principia* owed their genesis to Newton's unpublished invention of calculus.

Among his earliest work on the topic was a paper titled *On Analysis by Means of Equations with an Infinite Number of Terms*, which he circulated to a few friends in 1669. In modern terminology, he asked what the equation of a function f(x) is, if the area under its graph is of the form x^m. (Actually he asked something slightly more general, but let's keep it simple.) He deduced, to his own satisfaction, that the answer is $f(x) = mx^{m-1}$.

Newton's approach to calculating derivatives was much like that of Leibniz, except that he used *o* in place of dx, so his method suffers from the same logical problem: it seems to be only approximate. But Newton could show that by assuming *o* to be very small, the approximation would become ever better. In the limit, when *o* becomes as small as we please, the error vanishes. So, Newton maintained, his end result was *exact*. He introduced a new word, *fluxion*, to capture the main idea – that of a quantity flowing towards zero but not actually getting there.

In 1671 he wrote a more extensive treatment, the *Method of Fluxions and Infinite Series*. The first book on calculus was not published until 1711; the second appeared in 1736. It is clear that by 1671 Newton possessed most of the basic ideas of calculus.

Objectors to this procedure, notably Bishop George Berkeley in his 1734 book *The Analyst, a Discourse Addressed to an Infidel Mathematician*, pointed out that it is illogical to divide numerator and denominator by *o* when later *o* is set to 0. In effect, the procedure conceals the fact that the fraction is actually 0/0, which is well known to be meaningless. Newton responded that he was not actually setting *o* equal to 0; he was working out what happened when *o* became as close as we wish to 0 *without actually getting there*. The method was about fluxions, not numbers.

The mathematicians sought refuge in physical analogies – Leibniz referred to the 'spirit of finesse' as opposed to the 'spirit of logic' – but Berkeley was perfectly correct. It took more than a century to find a good answer to his objections, by defining the intuitive notion of 'passing to a limit' in a rigorous manner. Calculus then turned into a more subtle subject, *analysis*. But for a century after the invention of calculus, nobody except Berkeley worried much about its logical foundations, and calculus flourished despite this flaw.

It flourished because Newton was right, but it would take nearly 200 years before his concept of a fluxion was formulated in a logically acceptable way, in terms of limits. Fortunately for mathematics, progress was not halted until a decent logical foundation was discovered. Calculus was too useful, and too important, to be held up over a few logical quibbles. Berkeley was incensed, maintaining that the method seemed to work only because various errors cancelled each other out. He was right – but he failed to enquire *why* they always cancelled out. Because if that were the case, they weren't really errors at all!

Associated with differentiation is the reverse process, *integration*. The integral of f(x), written, $\int f(x)dx$, is whichever function yields f(x) when it is differentiated. Geometrically it represents the area under the graph of the function f. The *definite integral* $\int_a^b f(x)dx$ is the area under this graph between the values $x = a$ and $x = b$.

Derivatives and integrals solved problems that had taxed the ingenuity of previous mathematicians. Velocities, tangents, maxima and minima could all be found using differentiation. Lengths, areas and volumes could be calculated by integration. But there was more. Surprisingly, it seemed that the patterns of nature were written in the language of the calculus.

The English get left behind

As the importance of the calculus became ever clearer, greater prestige became attached to its creator. But who *was* the creator?

We have seen that Newton began thinking about calculus in 1665, but did not publish anything on the topic until 1687. Leibniz, whose ideas ran along roughly similar lines to Newton's, had started working on calculus in 1673, and published his first papers on the topic in 1684. The two worked independently, but

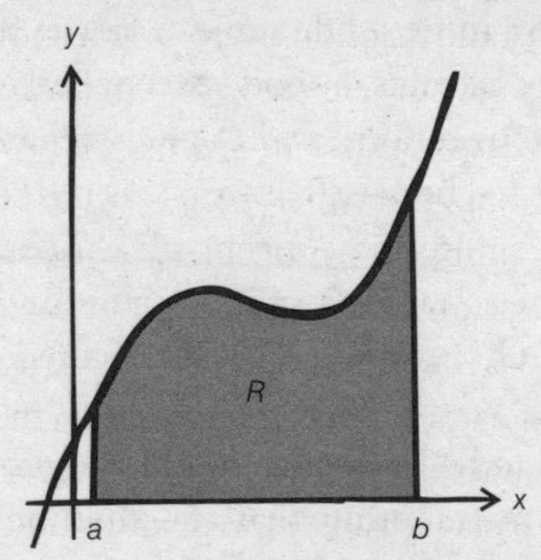

The definite integral

What calculus did for them

An early use of calculus to understand natural phenomena was the question of the shape of a hanging chain. The question was controversial; some mathematicians thought the answer was a parabola, others disagreed. In 1691 Leibniz, Christiaan Huygens and Joahnn Bernoulli all published proposed solutions. The clearest was Bernoulli's. He wrote down a differential equation to describe the position of the chain, based on Newtonian mechanics and Newton's laws of motion.

The solution, it turned out, was not a parabola, but a curve known as a catenary, with equation

$$y = k(e^x + e^{-x})$$

for constant k.

The cables of suspension bridges, however, are parabolic. The difference arises because these cables carry the weight of the bridge, as well as their own weight. Again, this can be demonstrated using calculus.

Leibniz *might* have learned about Newton's work when he visited Paris in 1672 and London in 1673; Newton had sent a copy of *On Analysis* to Barrow in 1669, and Leibniz talked to several people who also knew Barrow and so might have known about this work.

When Leibniz published his work in 1684, some of Newton's friends took umbrage – probably because Newton had been pipped to the publication post and they all belatedly realized what was at stake – and accused Leibniz of stealing Newton's ideas. The continental mathematicians, especially the Bernoullis, leaped to Leibniz's defence, suggesting that it was Newton, not Leibniz, who was guilty of plagiarism. In point of fact, both men had made their discoveries pretty much independently, as their unpublished manuscripts show; to muddy the waters, both had leaned heavily on previous work of Barrow, who probably had better grounds for grievance than either of them.

The accusations could easily have been withdrawn, but instead the dispute grew more heated; John Bernoulli extended his distaste from Newton to the entire English nation. The end result was a disaster for English mathematics, because the English doggedly stuck with Newton's geometric style of thinking, which was difficult to use, whereas the continental analysts employed Leibniz's more formal, algebraic methods, and pushed the subject ahead at a rapid pace. Most of the payoff in mathematical physics therefore went to the French, Germans, Swiss and Dutch, while English mathematics languished in a backwater.

The differential equation

The most important single idea to emerge from the flurry of work on calculus was the existence, and the utility, of a novel kind of equation – the *differential equation*. Algebraic equations relate various powers of an unknown number. Differential equations are much grander: they relate various derivatives of an unknown *function*.

Newton's laws of motion tell us that if $y(t)$ is the height of a particle moving under gravity near the Earth's surface, then the second derivative d^2y/dt^2 is proportional to the force g that acts; specifically,

$$g = m\frac{d^2y}{dt^2}$$

where m is the mass of the particle. This equation does not specify the function y directly. Instead, it specifies a property of its second derivative. We must solve the differential equation to find y itself. Two successive integrations yield the solution

$$y = \frac{gt^2}{2m} + at + b$$

What calculus does for us

Differential equations abound in science: they are by far the commonest way to model natural systems. To choose one application at random, they are used routinely to calculate the trajectories of space probes, such as the Mariner mission to Mars, the two Pioneer craft that explored the solar system and gave us such wonderful images of Jupiter, Saturn, Uranus and Neptune, and the recent Mars Rovers *Spirit* and *Opportunity*, six-wheeled robot vehicles that explored the Red Planet.

The Cassini mission, currently exploring Saturn and its moons, is another example. Among its discoveries is the existence of lakes of liquid methane and ethane on Saturn's moon Titan. Of course, calculus is not the only technique used by space missions – but without it, these missions would literally never have got off the ground.

More practically, every aircraft that flies, every car that travels the road and every suspension bridge and earthquake proof building owes its design in part to calculus. Even our understanding of how animal populations change size over time stems from differential equations. The same goes for the spread of epidemics, where calculus models are used to plan the most effective way to intervene and prevent disease spreading. A recent model of the foot and mouth disease epidemic in the UK has shown that the strategy adopted at the time was not the best available.

where b is the initial height of the particle and a is its initial velocity. The formula tells us that the graph of height y against time t is an upside-down parabola. This is Galileo's observation.

The pioneering efforts of Copernicus, Kepler, Galileo and other Renaissance scientists led to the discovery of mathematical patterns in the natural world. Some apparent patterns turned out to be

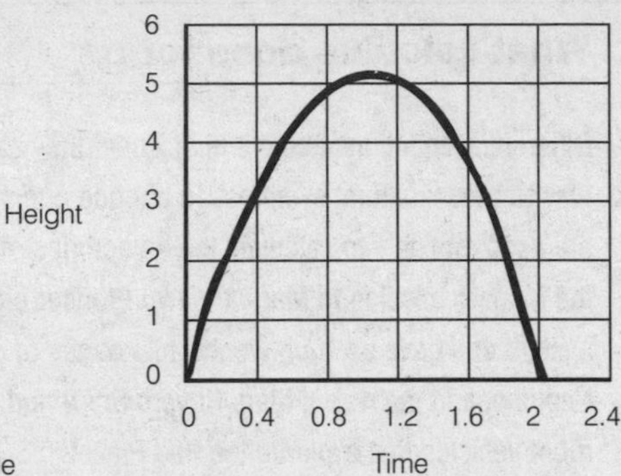

Parabolic trajectory of a projectile

spurious, and were discarded; others provided very accurate models of nature, and were retained and developed. From these early beginnings, the notion that we live in a 'clockwork universe', running according to rigid, unbreakable rules, emerged, despite serious religious opposition, mainly from the Church of Rome.

Newton's great discovery was that nature's patterns seem to manifest themselves not as regularities in certain quantities, but as relations among their *derivatives*. The laws of nature are written in the language of calculus; what matters are not the values of physical variables, but the rates at which they change. It was a profound insight, and it created a revolution, leading more or less directly to modern science, and changing our planet forever.

CHAPTER 9

Patterns in Nature

Formulating laws of physics

The main message in Newton's *Principia* was not the specific laws of nature that he discovered and used, but the idea that such laws exist – together with evidence that the way to model nature's laws mathematically is with differential equations. While England's mathematicians engaged in sterile vituperation over Leibniz's alleged (and totally fictitious) theft of Newton's ideas about calculus, the continental mathematicians were cashing in on Newton's great insight, making important inroads into celestial mechanics, elasticity, fluid dynamics, heat, light and sound – the core topics of mathematical physics. Many of the equations that they derived remain in use to this day, despite – or perhaps because of – the many advances in the physical sciences.

Differential equations

To begin with, mathematicians concentrated on finding explicit formulas for solutions of particular kinds of ordinary differential equation. In a way this was unfortunate, because formulas of this type usually fail to exist, so attention became focused on equations that could be solved by a formula rather than equations that genuinely described nature. A good example is the differential equation for a pendulum, which takes the form

$$\frac{d^2\theta}{dt^2} + k^2 \sin \theta = 0$$

for a suitable constant k, where t is time and θ is the angle at which the pendulum hangs, with $\theta = 0$ being vertically downwards. There is no solution of this equation in terms of classical functions (polynomial, exponential, trigonometric, logarithmic and so on). There does exist a solution using elliptic functions, invented more than a century later. However, if it is assumed that the angle is small, so we are considering a pendulum making small oscillations, then $\sin \theta$ is approximately equal to θ, and the smaller θ becomes the better this approximation is. So the differential equation can be replaced by

$$\frac{d^2\theta}{dt^2} + k^2 \theta = 0$$

and now there is a formula for the solution, in general,

$$\theta = A \sin kt + B \cos kt$$

for constants A and B, determined by the initial position and angular velocity of the pendulum.

This approach has some advantages: for instance, we can quickly deduce that the period of the pendulum – the time taken to complete one swing – is $2\pi/k$. The main disadvantage is that the solution fails when θ becomes sufficiently large (and here even 20° is large if we want an accurate answer). There is also a question of rigour: is it the case that an exact solution to an approximate equation yields an approximate solution to the exact equation? Here the answer is yes, but this was not proved until about 1900.

The second equation can be solved explicitly because it is *linear* – it involves only the first power of the unknown θ and its derivative, and the coefficients are constant. The prototype function for all linear differential equations is the exponential $y = e^x$. This satisfies the equation

$$\frac{dy}{dx} = y$$

That is, e^x is its own derivative. This property is one reason why the number e is so natural. A consequence is that the derivative of the natural logarithm $\log x$ is $1/x$, so the integral of $1/x$ is $\log x$. Any linear differential equation with constant coefficients can be solved using exponentials and trigonometric functions (which we will see are really exponentials in disguise).

Types of differential equation

There are two types of differential equation. An *ordinary* differential equation (ODE) refers to an unknown function y of a single variable x, and relates various derivatives of y, such as dy/dx and d^2y/dx^2. The differential equations described so far have been ordinary ones. Far more difficult, but central to mathematical physics, is the concept of a *partial* differential equation (PDE). Such an equation refers to an unknown function y of *two or more* variables, such as $f(x, y, t)$ where x and y are coordinates in the plane and t is time. The PDE relates this function to expressions in its partial derivatives with respect to each of the variables. A new notation is used to represent derivatives of some variables with respect to others, while the remainder are held fixed. Thus, $\partial x/\partial t$ indicates the rate of change of x with respect to time, while y is held constant. This is called a *partial derivative* – hence the term partial differential equation.

Euler introduced PDEs in 1734 and d'Alembert did some work on them in 1743, but these early investigations were isolated and special. The first big breakthrough came in 1746, when d'Alembert returned to an old problem, the vibrating violin string. Johann Bernoulli had discussed a finite element version of this question in

1727, considering the vibrations of a finite number of point masses spaced equally far apart along a weightless string. D'Alembert treated a continuous string, of uniform density, by applying Bernoulli's calculations to n masses, and then letting n tend to infinity. Thus, a continuous string was in effect thought of as infinitely many infinitesimal segments of string, connected together.

Starting from Bernoulli's results, which were based on Newton's law of motion, and making some simplifications (for example, that the size of the vibration is small), d'Alembert was led to the PDE

$$\frac{\partial^2 y}{\partial t^2} = a^2 \frac{\partial^2 y}{\partial x^2}$$

where $y = y(x,t)$ is the shape of the string at time t, as a function of the horizontal coordinate x. Here a is a constant related to the tension in the string and its density. By an ingenious argument, d'Alembert proved that the general solution of his PDE has the form

$$y(x, t) = f(x + at) + f(x - at)$$

where f is periodic, with period twice the length of the string, and f is an odd function – that is, $f(-z) = -f(z)$. This form satisfies the natural boundary condition that the ends of the string do not move.

Wave equation

We now call d'Alembert's PDE the *wave equation*, and interpret its solution as a superposition of symmetrically placed waves, one moving with velocity a and the other velocity $-a$ (that is, travelling in the opposite direction). It has become one of the most important equations in mathematical physics, because waves arise in many different circumstances.

Euler spotted d'Alembert's paper, and immediately tried to improve on it. In 1753 he showed that without the boundary conditions, the general solution is

$$y(x, t) = f(x + at) + g(x - at)$$

where f and g are periodic, but satisfy no other conditions. In particular, these functions can have different formulas in different ranges of x, a feature that Euler referred to as discontinuous functions, though in today's terminology they are continuous but have discontinuous first derivatives.

In an earlier paper published in 1749 he pointed out that (for simplicity we will take the length of the string to be 2π) the simplest odd periodic functions are trigonometric functions

$$f(x) = \sin x, \sin 2x, \sin 3x, \sin 4x, \ldots$$

and so on. These functions represent pure sinusoidal vibrations of frequencies 1, 2, 3, 4 and so on. The general solution, said Euler, is a superposition of such curves. The basic sine curve sin x is the fundamental mode of vibration, and the others are higher modes – what we now call harmonics.

Comparison of Euler's solution of the wave equation with d'Alembert's led to a foundational crisis. D'Alembert did not

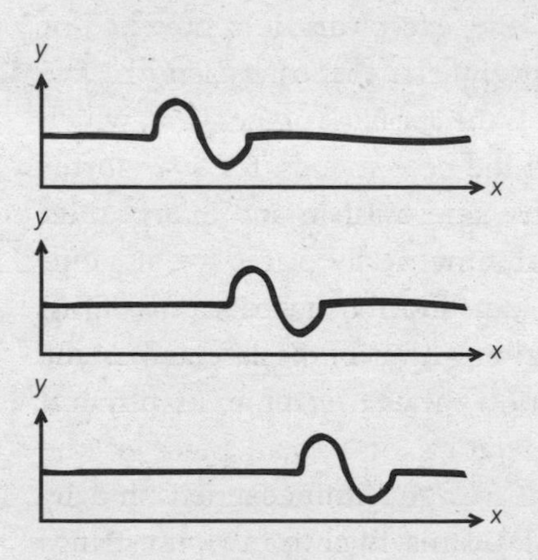

Successive snapshots of a wave travelling from left to right

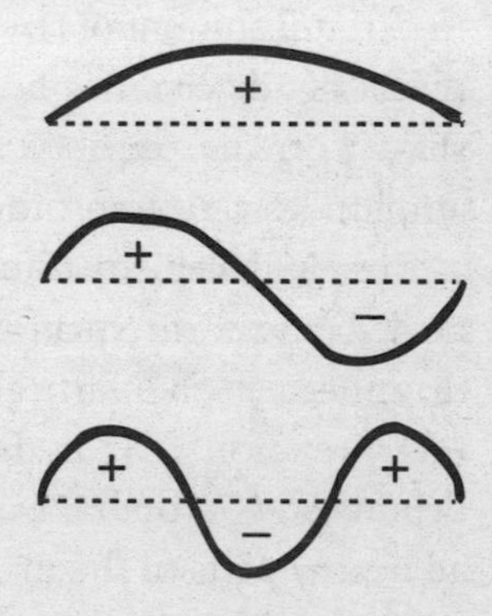

Vibrational modes of a string

recognize the possibility of discontinuous functions in Euler's sense. Moreover, there seemed to be a fundamental flaw in Euler's work, because trigonometric functions are continuous, and so are all (finite) superpositions of them. Euler had not committed himself on the issue of finite superpositions versus infinite ones – in those days no one was really very rigorous about such matters, having yet to learn the hard way that it matters. Now the failure to make such a distinction was causing serious trouble. The controversy simmered until later work by Fourier caused it to boil over.

Music, light, sound and electromagnetism

The ancient Greeks knew that a vibrating string can produce many different musical notes, depending on the position of the nodes, or rest points. For the fundamental frequency, only the end points are at rest. If the string has a node at its centre, then it produces a note one octave higher; and the more nodes there are, the higher the frequency of the note will be. The higher vibrations are called *overtones* or *harmonics*.

The vibrations of a violin string are *standing waves* – the shape of the string at any instant is the same, except that it is stretched or compressed in the direction at right angles to its length. The maximum amount of stretching is the *amplitude* of the wave, which physically determines how loud the note sounds. The waveforms shown on the previous page are sinusoidal in shape; and their amplitudes vary sinusoidally with time.

In 1759 Euler extended these ideas from strings to drums. Again he derived a wave equation, describing how the displacement of the drumhead in the vertical direction varies over time. Its physical interpretation is that the acceleration of a small piece of the drumhead is proportional to the average tension exerted on it by all nearby parts of the drumhead. Drums differ from violin strings not only in their dimensionality – a drum is a flat two-dimensional membrane – but in having a much more interesting *boundary*. In this

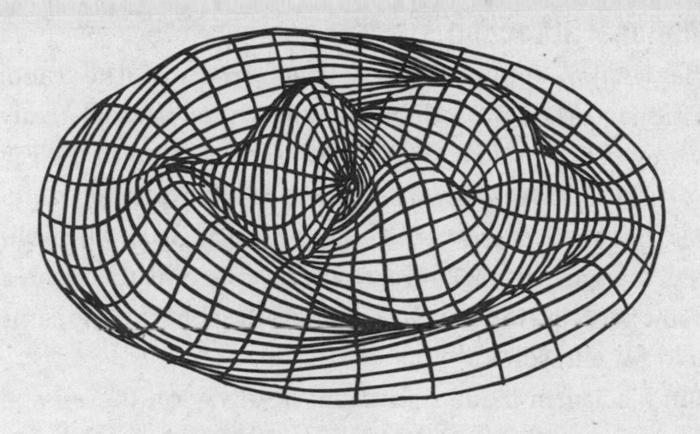

Vibrations of a circular drumhead

whole subject, boundaries are absolutely crucial. The boundary of a drum can be any closed curve, and the key condition is that the boundary of the drum is *fixed*. The rest of the drumhead can move, but its rim is firmly strapped down.

The mathematicians of the 18th century were able to solve the equations for the motion of drums of various shapes. Again they found that all vibrations can be built up from simpler ones, and that these yield a specific list of frequencies. The simplest case is the rectangular drum, the simplest vibrations of which are combinations of sinusoidal ripples in the two perpendicular directions. A more difficult case is the circular drum, which leads to new functions called *Bessel functions*. The amplitudes of these waves still vary sinusoidally with time, but their spatial structure is more complicated.

The wave equation is exceedingly important. Waves arise not only in musical instruments, but in the physics of light and sound. Euler found a three-dimensional version of the wave equation, which he applied to sound waves. Roughly a century later, James Clerk Maxwell extracted the same mathematical expression from his equations for electromagnetism, and predicted the existence of radio waves.

Gravitational attraction

Another major application of PDEs arose in the theory of gravitational attraction, otherwise known as *potential theory*. The motivating problem was the gravitational attraction of the Earth, or of any other planet. Newton had modelled planets as perfect spheres, but their true form is closer to that of an ellipsoid. And whereas the gravitational attraction of a sphere is the same as that of a point particle (for distances outside the sphere), the same does not hold for ellipsoids.

Colin Maclaurin made significant headway on these issues in a prize-winning memoir of 1740 and a subsequent book *Treatise of Fluxions* published in 1742. His first step was to prove that if a fluid of uniform density spins at a uniform speed, under the influence of its own gravity, then the equilibrium shape is an oblate spheroid – an ellipsoid of revolution. He then studied the attractive forces generated by such a spheroid, with limited success. His main result was that if two spheroids have the same foci, and if a particle lies either on the equatorial plane or the axis of revolution, then the force exerted on it by either spheroid is proportional to their masses.

In 1743 Clairaut continued working on this problem in his *Théorie de la Figure de la Terre*. But the big breakthrough was made by Legendre. He proved a basic property of not just spheroids, but any solid of revolution: if you know its gravitational attraction at every point along its axis, then you can deduce the attraction at any other point. His method was to express the attraction as an integral in spherical polar coordinates. Manipulating this integral, he expressed its value as a superposition of spherical harmonics. These are determined by special functions now called *Legendre polynomials*. In 1784, he pursued this topic, proving many basic properties of these polynomials.

The fundamental PDE for potential theory is Laplace's equation, which appears in Laplace's five-volume *Traité de Mécanique Céleste*

An ellipsoid

(*Treatise on Celestial Mechanics*) published from 1799 onwards. It was known to earlier researchers, but Laplace's treatment was definitive. The equation takes the form

$$\frac{\partial^2 V}{\partial x^2} + \frac{\partial^2 V}{\partial y^2} + \frac{\partial^2 V}{\partial z^2} = 0$$

where $V(x, y, z)$ is the potential at a point (x, y, z) in space. Intuitively, it says that the value of the potential at any given point is the average of its values over a tiny sphere surrounding that point. The equation is valid outside the body: inside, it must be modified to what is now known as Poisson's equation.

Heat and temperature

Successes with sound and gravitation encouraged mathematicians to turn their attention to other physical phenomena. One of the most significant was heat. At the start of the 19th century the science of heat flow was becoming a highly practical topic, mainly because of the needs of the metalworking industry, but also because of a growing interest in the structure of the Earth's interior, and in particular the temperature inside the planet. There is no direct way to measure the temperature a thousand miles or more below the Earth's surface, so the only available methods were indirect, and an understanding of how heat flowed through bodies of various compositions was essential.

In 1807 Joseph Fourier submitted a paper on heat flow to the French Academy of Sciences, but the referees rejected it because it was insufficiently developed. To encourage Fourier to continue the work, the Academy made heat flow the subject of its 1812 grand prize. The prize topic was announced well ahead of time, and by 1811 Fourier had revised his ideas, put them in for the prize, and won. However, his work was widely criticized for its lack of logical rigour and the Academy refused to publish it as a memoir. Fourier, irritated by this lack of appreciation, wrote his own book, *Théorie Analytique de la Chaleur* (*Analytic Theory of Heat*), published in 1822. Much of the 1811 paper was included unchanged, but there was extra material too. In 1824 Fourier got even: he was made Secretary of the Academy, and promptly published his 1811 paper as a memoir.

Fourier's first step was to derive a PDE for heat flow. With various simplifying assumptions – the body must be homogeneous (with the same properties everywhere) and isotropic (no direction within it should behave differently from any other), and so on, he came up with what we now call the *heat equation*, which describes how the temperature at any point in a three-dimensional body varies with time. The heat equation is very similar in form to Laplace's equation and the wave equation, but the partial derivative with respect to time is of first order, not second. This tiny change makes a huge difference to the mathematics of the PDE.

There were similar equations for bodies in one and two dimensions (rods and sheets) obtained by removing the terms in z (for two dimensions) and then y (for one). Fourier solved the heat equation for a rod (whose length we take to be π), the ends of which are maintained at a fixed temperature, by assuming that at time $t = 0$ (the initial condition) the temperature at a point x on the rod is of the form

$$b_1 \sin x + b_2 \sin 2x + b_3 \sin 3x + \cdots$$

(an expression suggested by preliminary calculations) and deduced that the temperature must then be given by a similar but more complicated expression in which each term is multiplied by a suitable exponential function. The analogy with harmonics in the wave equation is striking. But there each mode given by a pure sine function oscillates indefinitely without losing amplitude, whereas here each sinusoidal mode of the temperature distribution decays exponentially with time, and the higher modes decay more rapidly.

The physical reason for the difference is that in the wave equation energy is conserved, so the vibrations cannot die down. But in the heat equation, the temperature *diffuses* throughout the rod, and is lost at the ends because these are kept cool.

The upshot of Fourier's work is that whenever we can expand the initial temperature distribution in a **Fourier series** – a series of sine and cosine functions like the one above – then we can immediately read off how the heat flows through the body as time passes. Fourier considered it obvious that any initial distribution of temperature could be so expressed, and this is where the trouble began, because a few of his contemporaries had been worrying about precisely this issue for some time, in connection with waves, and had convinced themselves that it was much harder than it seemed.

Fourier's argument for the existence of an expansion in sines and cosines was complicated, confused and wildly non-rigorous. He went all round the mathematical houses to derive, eventually, a simple expression for the coefficients b_1, b_2, b_3, etc. Writing $f(x)$ for the initial temperature distribution, his result was

$$b_n = \frac{2}{\pi} \int_0^{\pi} f(u) \sin(nu)\, du$$

Euler had already written down this formula in 1777, in the

How Fourier Series Work

A typical discontinuous function is the *square wave S(x)*, which takes the values 1 when $-\pi < x \leq 0$ and -1 when $0 < x \leq \pi$, and has period 2π. Applying Fourier's formula to the square wave we get the series

$$S(x) = \sin x + \frac{1}{3}\sin 3x + \frac{1}{5}\sin 5x + \cdots$$

The terms add up, as shown in the diagram below.

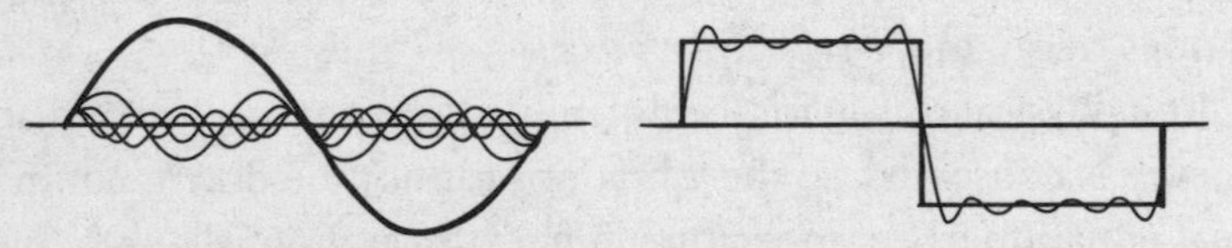

Fourier expansion of a square wave: left, the component sine curves; right, their sum

Although the square wave is discontinuous, every approximation is continuous. However, the wiggles build up as more and more terms are added, making the graph of the Fourier series become increasingly steep near the discontinuities. This is how an infinite series of continuous functions can develop a discontinuity.

context of the wave equation for sound, and he proved it using the clever observation that distinct modes, sin mx and sin nx, are *orthogonal*, meaning that

$$\int_0^\pi \sin(mx)\sin(nx)\,dx$$

is zero whenever m and n are distinct integers, but non-zero – in fact, equal to $\pi/2$ – when $m = n$. If we assume that $f(x)$ has a

Fourier expansion, multiply both sides by sin nx and integrate, then every term except one disappears, and the remaining terms yield Fourier's formula for b_n.

Fluid dynamics

No discussion of the PDEs of mathematical physics would be complete without mentioning fluid dynamics. Indeed, this is an area of enormous practical significance, because these equations describe the flow of water past submarines, of air past aircraft, and even the flow of air past Formula 1 racing cars.

Euler started the subject in 1757 by deriving a PDE for the flow of a fluid of zero viscosity, that is zero 'stickiness'. This equation remains

What differential equations did for them

Kepler's model of elliptical orbits is not exact. It would be if there were only two bodies in the solar system, but when a third body is present, it changes (perturbs) the elliptical orbit. Because the planets are spaced quite a long way apart, this problem affects the detailed motion, and most orbits remain close to ellipses. However, Jupiter and Saturn behave strangely, sometimes lagging behind where they ought to be and sometimes pulling ahead. This effect is caused by their mutual gravitation, together with that of the Sun.

Newton's law of gravitation applies to any number of bodies, but the calculations become very difficult when there are three bodies or more. In 1748, 1750 and 1752 the French Academy of Sciences offered prizes for accurate calculations of the movements of Jupiter and Saturn. In 1748 Euler used differential equations to study how Jupiter's gravity perturbs the orbit of Saturn, and won the prize. He tried again in 1752 but his work contained significant mistakes. However, the underlying ideas later turned out to be useful.

Sofia Vasilyevna Kovalevskaya

1850–1891

Sofia Kovalevskaya was the daughter of an artillery general and a member of the Russian nobility. It so happened that her nursery walls had been papered with pages from lecture notes on analysis. At the age of 11 she took a close look at her wallpaper and taught herself calculus. She became attracted to mathematics, preferring it to all other areas of study. Her father tried to stop her, but she carried on regardless, reading an algebra book when her parents were sleeping.

In order to travel and obtain an education, she was obliged to marry, but the marriage was never a success. In 1869 she studied mathematics in Heidelberg, but because female students were not permitted, she had to persuade the university to let her attend lectures on an unofficial basis. She showed impressive mathematical talent, and in 1871 she went to Berlin, where she studied under the great analyst Karl Weierstrass. Again, she was not permitted to be an official student, but Weierstrass gave her private lessons.

She carried out original research, and by 1874 Weierstrass said that her work was suitable for a doctorate. She had written three papers, on PDEs, elliptic functions and the rings of Saturn. In the same year Göttingen University awarded her a doctoral degree. The PDE paper was published in 1875.

In 1878 she had a daughter, but returned to mathematics in 1880, working on the refraction of light. In 1883 her husband, from whom she had separated, committed suicide, and she spent more and more time working on her mathematics to assuage her feelings of guilt. She obtained a university position in Stockholm, giving lectures in 1884. In 1889 she became the third female professor ever at a European university, after Maria Agnesi (who never took up her post) and the physicist Laura Bassi. Here she did research on the motion of a rigid body, entered it for a prize offered by the Academy of Sciences in 1886, and won. The jury found the work so brilliant that they increased the prize money. Subsequent work on the same topic was awarded a prize by the Swedish Academy of Sciences, and led to her being elected to the Imperial Academy of Sciences.

realistic for some fluids, but it is too simple to be of much practical use. Equations for a viscous fluid were derived by Claude Navier in 1821, and again by Poisson in 1829. They involve various partial derivatives of the fluid velocity. In 1845 George Gabriel Stokes deduced the same equations from more basic physical principles, and they are therefore known as the *Navier–Stokes equations*.

Ordinary differential equations

We close this section with two far-reaching contributions to the use of ODEs (ordinary differential equations) in mechanics. In 1788 Lagrange published his *Mécanique Analytique* (*Analytical Mechanics*), proudly pointing out that

> 'One will not find figures in this work. The methods that I expound require neither constructions, nor geometrical or mechanical arguments, but only algebraic operations, subject to a regular and uniform course.'

At that period, the pitfalls of pictorial arguments had become apparent, and Lagrange was determined to avoid them. Pictures are now back in vogue, though supported by solid logic, but Lagrange's insistence on a formal treatment of mechanics inspired a new unification of the subject, in terms of generalized coordinates. Any

What differential equations do for us

There is a direct link between the wave equation and radio and television. Around 1830 Michael Faraday carried out experiments on electricity and magnetism, investigating the creation of a magnetic field by an electric current, and an electric field by a moving magnet. Today's dynamos and electric motors are direct descendants from his apparatus. In 1864 James Clerk Maxwell reformulated Faraday's theories as mathematical equations for electromagnetism: *Maxwell's equations*. These are PDEs involving the magnetic and electric fields.

A simple deduction from Maxwell's equations leads to the wave equation. This calculation shows that electricity and magnetism can travel together like a wave, with the speed of light. What travels at the speed of light? Light. So light is an electromagnetic wave. The equation placed no limitations on the frequency of the wave, and light waves occupy a relatively small range of frequencies, so physicists deduced that there

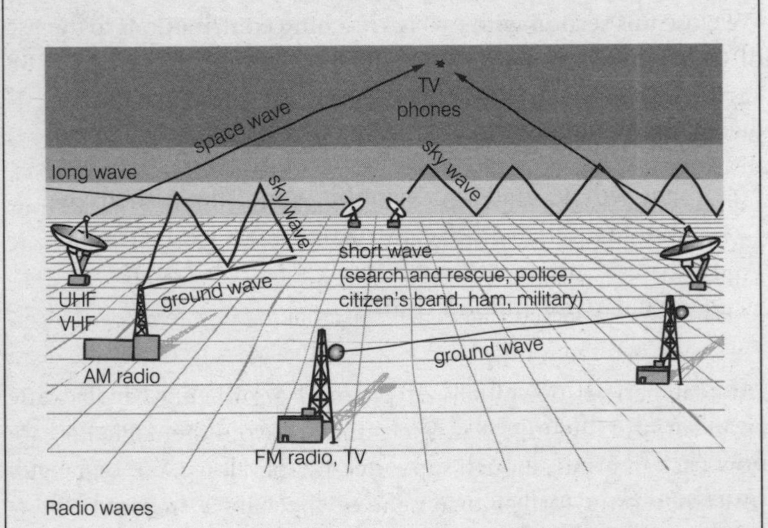

Radio waves

ought to be electromagnetic waves with other frequencies. Heinrich Hertz demonstrated the physical existence of such waves, and Guglielmo Marconi turned them into a practical device: radio. The technology snowballed. Television and radar also rely on electro-magnetic waves. So do GPS satellite navigation, mobile phones and wireless computer communications.

system can be described using many different variables. For a pendulum, for instance, the usual coordinate is the angle at which the pendulum is hanging, but the horizontal distance between the bob and the vertical would do equally well.

The equations of motion look very different in different coordinate systems, and Lagrange felt this was inelegant. He found a way to rewrite the equations of motion in a form that looks the same in every coordinate system. The first innovation is to pair off the coordinates: to every position coordinate q (such as the angle of the pendulum) there is associated the corresponding velocity coordinate, $\dot{q}$(the rate of angular motion of the pendulum). If there are k position coordinates, there are also k velocity coordinates. Instead of a second-order differential equation in the positions, Lagrange derived a first-order differential equation in the positions and the velocities. He formulated this in terms of a quantity now called the *Lagrangian*.

Hamilton improved Lagrange's idea, making it even more elegant. Physically, he used momentum instead of velocity to define the extra coordinates. Mathematically, he defined a quantity now called the *Hamiltonian*, which can be interpreted – for many systems – as energy. Theoretical work in mechanics generally uses the Hamiltonian formalism, which has been extended to quantum mechanics as well.

Physics goes mathematical

Newton's *Principia* was impressive, with its revelation of deep mathematical laws underlying natural phenomena. But what happened next was even more impressive. Mathematicians tackled the entire panoply of physics – sound, light, heat, fluid flow, gravitation, electricity, magnetism. In every case, they came up with differential equations that described the physics, often very accurately.

The long-term implications have been remarkable. Many of the most important technological advances, such as radio, television and commercial jet aircraft depend, in numerous ways, on the mathematics of differential equations. The topic is still the subject of intense research activity, with new applications emerging almost daily. It is fair to say that Newton's invention of differential equations, fleshed out by his successors in the 18th and 19th centuries, is in many ways responsible for the society in which we now live. This only goes to show what is lurking just behind the scenes, if you care to look.

CHAPTER 10

Impossible Quantities

Can negative numbers have square roots?

Mathematicians distinguish several different kinds of number, with different properties. What really matters is not the individual numbers, but the system to which they belong – the company they keep.

Four of these number systems are familiar: the *natural numbers*, 1, 2, 3, ...; the *integers*, which also include zero and negative whole numbers; the *rational numbers*, composed of fractions p/q where p and q are integers and q is not zero; and the *real numbers*, generally introduced as decimals that can go on forever – whatever that means – and represent both the rational numbers, as repeating decimals, and irrational numbers like $\sqrt{2}$, e and π whose decimal expansions do not ever repeat the same block of digits.

Integers

The name integer just means whole; the other names give the impression that the systems concerned are sensible, reasonable things – natural, rational and of course real. These names reflect, and encourage, a long-prevailing view that numbers are features of the world around us.

Many people think that the only way you can do mathematical research is to invent new numbers. This view is almost always wrong; a lot of mathematics is not about numbers at all, and in any case the usual aim is to invent new theorems, not new numbers. Occasionally, however, 'new numbers' do arise. And one such invention, a so-called 'impossible' or 'imaginary' number, completely changed the face of mathematics, adding immeasurably to its power. That number was the square root of minus one. To early mathematicians, such a description seemed ridiculous, because the square of any number is always positive. So, negative numbers cannot have square roots.

But just suppose they *did*. What would happen?

It took mathematicians a long time to appreciate that numbers are artificial inventions made by human beings; very effective inventions for capturing many aspects of nature, to be sure, but no more a *part* of nature than one of Euclid's triangles or a formula in calculus. Historically, we first see mathematicians starting to struggle with this philosophical question when they began to realize that imaginary numbers were inevitable, useful and somehow on a par with the more familiar real ones.

Problems with cubics

Revolutionary mathematical ideas are seldom discovered in the simplest and (with hindsight) most obvious context. They almost always emerge from something far more complicated. So it was with the square root of minus one. Nowadays, we normally introduce this number in terms of the quadratic equation $x^2 + 1 = 0$, the solution of which is the square root of minus one – whatever that means. Among the first mathematicians to wonder whether it had a sensible meaning were the Renaissance algebraists, who ran into square roots of negative numbers in a surprisingly indirect way: the solution of cubic equations.

Recall that del Ferro and Tartaglia discovered algebraic solutions to cubic equations, later written up by Cardano in his *Ars Magna*. In modern symbols, the solution of a cubic equation $x^3 + ax = b$ is

$$x = \sqrt[3]{\frac{b}{2} + \sqrt{\frac{a^3}{27} + \frac{b^2}{4}}} + \sqrt[3]{\frac{b}{2} - \sqrt{\frac{a^3}{27} + \frac{b^2}{4}}}$$

The Renaissance mathematicians expressed this solution in words and abbreviations, but the procedure was the same.

Sometimes this formula worked beautifully, but sometimes it ran into trouble. Cardano noticed that when the formula is applied to the equation $x^3 = 15x + 4$, with the obvious solution $x = 4$, the result is expressed as

$$x = \sqrt[3]{2 + \sqrt{-121}} + \sqrt[3]{2 - \sqrt{-121}}$$

This expression seemed to have no sensible meaning, however, because -121 has no square root. A puzzled Cardano wrote to Tartaglia, asking for clarification, but Tartaglia missed the point and his reply was distinctly unhelpful.

An answer of sorts was provided by Rafael Bombelli in his three-volume book *L'Algebra*, printed in Venice in 1572 and Bologna in 1579. Bombelli was worried that Cardano's *Ars Magna* was rather obscure, and he set out to write something clearer. He operated on the troublesome square root as if it were an ordinary number, noticing that

$$(2 + \sqrt{-1})^3 = 2 + \sqrt{-121}$$

and deducing the curious formula

$$\sqrt[3]{2 + \sqrt{-121}} = 2 + \sqrt{-1}$$

Similarly, Bombelli obtained the formula

$$\sqrt[3]{2 + \sqrt{-121}} = 2 - \sqrt{-1}$$

Now he could rewrite the sum of the two cube roots as

$$(2 + \sqrt{-1}) + (2 - \sqrt{-1}) = 4$$

So this strange method yielded the right answer, a perfectly normal integer, but it got there by manipulating 'impossible' quantities.

This was all very interesting, but *why did it work?*

Imaginary numbers

To answer this question, mathematicians had to develop good ways to think about square roots of negative quantities, and do calculations with them. Early writers, among them Descartes and Newton, interpreted these imaginary numbers as a sign that a problem has no solutions. If you wanted to find a number whose square was minus one, the formal solution, square root of minus one, was imaginary, so no solution existed. But Bombelli's calculation implied that there was more to imaginaries than that. They could be used to *find* solutions; they could occur when solutions *did* exist.

In 1673 John Wallis invented a simple way to represent imaginary numbers as points in a plane. He started from the familiar representation of real numbers as a line, with the positive numbers on the right and the negative ones on the left.

Then he introduced another line, at right angles to the first, and along this new line he placed the imaginaries.

This is like Descartes's algebraic approach to plane geometry, using coordinate axes. Real numbers form one axis in the picture, imaginaries another. Wallis did not state the idea in quite this form – his version was closer to Fermat's approach to coordinates than Descartes's. But the underlying point is the same. The remainder of the plane corresponds to complex numbers, which consist of two parts: one real, one imaginary. In Cartesian coordinates, we measure the real part along the real line and measure the imaginary part parallel to the imaginary line. So $3 + 2i$ lies 3 units to the right of the origin and 2 units up.

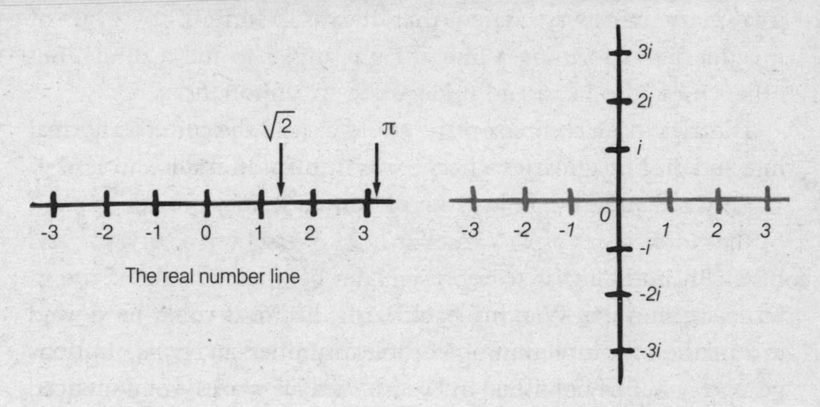

The real number line

Two copies of the real number line, placed at right angles

Wallis's idea solved the problem of giving meaning to imaginary numbers, but no one took the slightest notice. However, his idea slowly gained ground subconsciously. Most mathematicians stopped worrying that the square root of minus one could not occupy any position on the real line, and realized that it could live somewhere in the wider world of the complex plane. Some failed to appreciate the idea: in 1758 François Daviet de Foncenex, in a paper about

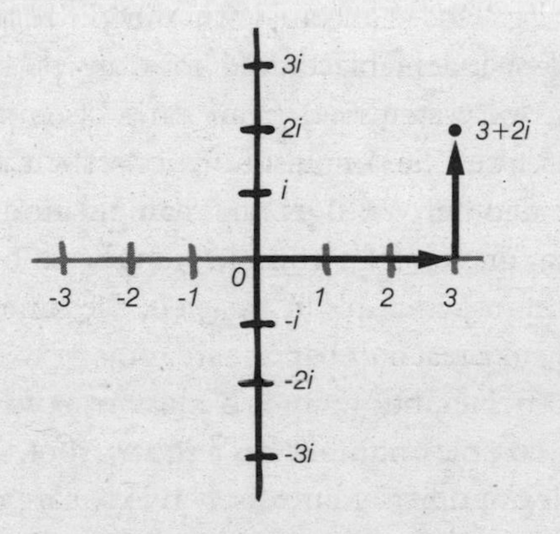

The complex plane according to Wessel

imaginary numbers, stated that it was pointless to think of imaginaries as forming a line at right angles to the real line. But others took it to heart and understood its importance.

The idea that a complex plane could extend the comfortable real line and give imaginaries a home was implicit in Wallis's work, but slightly obscured by the way he presented it. It was made explicit by the Norwegian Caspar Wessel in 1797. Wessel was a surveyor, and his main interest was to represent the geometry of the plane in terms of numbers. Working backwards, his ideas could be viewed as a method for representing complex numbers in terms of planar geometry. But he published in Danish, and his work went unnoticed until a century later, when it was translated into French. The French mathematician Jean-Robert Argand independently published the same representation of complex numbers in 1806, and Gauss discovered it independently of them both in 1811.

Complex analysis

If complex numbers had been good only for algebra, they might have remained an intellectual curiosity, of little interest outside pure mathematics. But as interest in calculus grew, and it took on a more rigorous form as analysis, people began to notice that a really interesting fusion of real analysis with complex numbers – *complex analysis* – was not only possible, but desirable. Indeed, for many problems, essential.

This discovery stemmed from early attempts to think about complex *functions*. The simplest functions, such as the square or the cube, depended only on algebraic manipulations, so it was easy to define these functions for complex numbers. To square a complex number, you just multiply it by itself, the same process that you would apply to a real number. Square roots of complex numbers are marginally trickier, but there is a pleasant reward for making the effort: *every* complex number has a square root. Indeed, every non-zero complex number has precisely two square roots, one equal to minus the other. So not only did augmenting the real numbers with a new number, i, provide -1 with a square root, it provided square roots for everything in the enlarged system of complex numbers.

What about sines, cosines, the exponential function and the logarithm? At this stage, things started to get very interesting, but also very puzzling, especially when it came to logarithms.

Like i itself, logarithms of complex numbers turned up in purely real problems. In 1702 Johann Bernoulli was investigating the process of integration, applied to reciprocals of quadratics. He knew a clever technique to carry out this task whenever the quadratic equation concerned has two real solutions, r and s. Then we can rewrite the expression to be integrated in terms of 'partial fractions'

$$\frac{1}{ax^2 + bx + c} = \frac{A}{x - r} + \frac{B}{x - s}$$

which leads to the integral

$$A \log (x - r) + B \log (x - s)$$

But what if the quadratic has no real roots? How can you integrate the reciprocal of $x^2 + 1$, for instance? Bernoulli realized that once you have defined complex algebra, the partial fraction trick still works, but now r and s are complex numbers. So, for example,

$$\frac{1}{x^2 + 1} = \frac{i/2}{x + i} + \frac{i/2}{x - i}$$

and the integral of this function takes the form

$$i/2 \log (x + i) + i/2 \log (x - i)$$

This final step was not fully satisfactory, because it demanded a definition of the logarithm of a complex number. Was it possible to make sense of such a statement?

Bernoulli thought it was, and proceeded to use his new idea to excellent effect. Leibniz also exploited this kind of thinking. But the mathematical details were not straightforward. By 1712 the two of them were arguing about a very basic feature of this approach. Forget *complex* numbers – what was the logarithm of a *negative real* number? Bernoulli thought that the logarithm of a negative real number should be real; Leibniz insisted that it was complex. Bernoulli had a kind of proof for his claim: assuming the usual formalism of calculus, the equation

$$\frac{d(-x)}{-x} = \frac{dx}{x}$$

can be integrated to yield

$$\log(-x) = \log(x)$$

However, Leibniz was unconvinced, and believed that the integration was correct only for positive real x.

This particular controversy was sorted out by Euler in 1749, and Leibniz was right. Bernoulli, said Euler, had forgotten that any integration involves an arbitrary constant. What Bernoulli should have deduced was that

$$\log(-x) = \log(x) + c$$

for some constant c. What was this constant? If logarithms of negative (and complex) numbers are to behave like logarithms of positive real numbers, which is the point of the whole game, then it should be true that

$$\log(-x) = \log(-1 \times x) = \log(-1) + \log x$$

so that $c = \log(-1)$. Euler then embarked on a series of beautiful calculations that produced a more explicit form for c. First, he found a way to manipulate various formulas involving complex numbers, assuming they behaved much like real numbers, and deduced a relation between trigonometric functions and the exponential:

$$e^{i\theta} = \cos \theta + i \sin \theta$$

a formula that had been anticipated in 1714 by Roger Cotes. Putting $\theta = \pi$, Euler obtained the delightful result that

$$e^{i\pi} = -1$$

relating the two fundamental mathematical constants e and π. It is remarkable that any such relation should exist, and even more remarkable that it is so simple. This formula regularly tops league tables for the 'most beautiful formula of all time'.

Taking the logarithm, we immediately deduce that

$$\log (-1) = i\pi$$

revealing the secret of that enigmatic constant c above: it is $i\pi$. As such, it is imaginary, so Leibniz was right and Bernoulli was wrong.

However, there is more, and it opens Pandora's box. If we put $\theta = 2\pi$, then

$$e^{2i\pi} = 1$$

So $\log (1) = 2i\pi$. Then the equation $x = x \times 1$ implies that

$$\log x = \log x + 2i\pi$$

from which we conclude that if n is any integer whatever,

$$\log x = \log x + 2ni\pi$$

At first sight, this makes no sense – it seems to imply that $2ni\pi = 0$ for all n. But there is a way to interpret it that does make sense. Over the complex numbers, the logarithmic function is many-valued. Indeed, unless the complex number z is zero, the function $\log z$ can take infinitely many distinct values. (When $z = 0$, the value $\log 0$ is not defined.)

Mathematicians were used to functions that could take several distinct values, the square root being the most obvious example: here, even a real number possessed two distinct square roots, one

What complex numbers did for them

The real and imaginary parts of a complex function satisfy the Cauchy–Riemann equations, which are closely related to the PDEs for gravitation, electricity, magnetism and some types of fluid flow in the plane. This connection made it possible to solve many equations of mathematical physics – but only for two-dimensional systems.

positive and the other negative. But *infinitely many* values? This was very strange.

Cauchy's theorem

What really put the cat among the pigeons was the discovery that you could do calculus – analysis – with complex functions, and that the resulting theory was elegant and *useful*. So useful, in fact, that the logical basis of the idea ceased to be an important issue. When something *works*, and you feel that you need it, you generally stop asking whether it makes sense.

The introduction of complex analysis seems to have been a conscious decision by the mathematical community – a generalization so obvious and compelling that any mathematician with any kind of sensitivity would want to see what transpired. In 1811 Gauss wrote a letter to a friend, the astronomer Friedrich Bessel, revealing his representation of complex numbers as points in a plane; he also mentioned some deeper results. Among them is a basic theorem upon which the whole of complex analysis hangs. Today we call it Cauchy's Theorem, because it was published by Cauchy, but Gauss had the idea much earlier in his unpublished writings.

This theorem concerns definite integrals of complex functions: that is, expressions

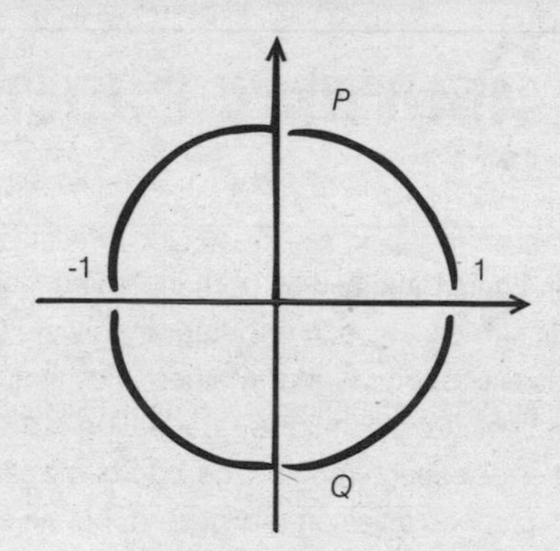

Two distinct paths *P* and *Q* from –1 to 1 in the complex plane

$$\int_a^b f(z)\,dz$$

where a and b are complex numbers. In real analysis this expression can be evaluated by finding an antiderivative $F(z)$ of $f(z)$, that is, a function $F(z)$ such that its derivative $dF(z)/dz = f(z)$. Then the definite integral is equal to $F(b) - F(a)$. In particular, its value depends only on the end points a and b, not on how you move from one to the other.

Complex analysis, said Gauss, is different. Now the value of the integral may depend on the *path* that the variable z takes as it moves from a to b. Because the complex numbers form a plane, their geometry is richer than that of the real line, and this is where that extra richness matters.

For example, suppose you integrate $f(z) = 1/z$ from $a = -1$ to $b = 1$. If the path concerned is a semicircle P lying above the real axis, then the integral turns out to be $-\pi i$. But if the path is a semicircle Q lying below the real axis, then the integral turns out to be πi. The two values are different, and the difference is $2\pi i$.

This difference, said Gauss, occurs because the function $1/z$ is badly behaved. It becomes infinite inside the region enclosed by the

Augustin-Louis Cauchy

1789–1857

Augustin-Louis Cauchy was born in Paris during a time of political turmoil. Laplace and Lagrange were family friends, so Cauchy was exposed to higher mathematics at an early age. He went to the École Polytechnique, graduating in 1807. In 1810 he carried out engineering work in Cherbourg, preparing for Napoleon's planned invasion of England, but he continued thinking about mathematics, reading Laplace's *Mécanique Céleste* (*Celestial Mechanics*) and Lagrange's *Théorie des Fonctions* (*Theory of Functions*).

He continually sought academic positions, without success, but kept working on mathematics. His famous paper on complex integrals, which effectively founded complex analysis, appeared in 1814, and he finally achieved his goal of an academic post, becoming assistant professor of analysis at the École Polytechnique a year later. His mathematics now flourished, and a paper on waves won him the 1816 prize of the Academy of Sciences. He continued to develop complex analysis, and in his 1829 *Leçons sur le Calcul Différentiel* he gave the first explicit definition of a complex function.

After the revolution of 1830 Cauchy briefly went to Switzerland, and in 1831 he became professor of theoretical physics in Turin. His courses were reported as being highly disorganized. By 1833 he was in Prague tutoring the grandson of Charles X, but the prince disliked both mathematics and physics, and Cauchy often lost his temper. He returned to Paris in 1838, regaining his post at the Academy but did not regain his teaching positions until Louis Philippe was deposed in 1848. Altogether, he published an astonishing 789 research articles on mathematics.

two paths. Namely, at $z = 0$, which here is the centre of the circle formed by the two paths. 'But if this does not happen ... I affirm,'

Gauss wrote to Bessel, 'that the integral has only one value even if taken over different paths provided [the function] does not become infinite in the space enclosed by the two paths. This is a very beautiful theorem, whose proof I shall give on a convenient occasion'. But he never did.

Instead, the theorem was rediscovered, and published by Augustin-Louis Cauchy, the true founder of complex analysis. Gauss may have had the ideas, but ideas are useless if no one gets to see them. Cauchy published his work. In fact, Cauchy seldom stopped publishing. It is said that the rule, still in force today, that the journal *Comptes Rendus de l'Academie Française* accepts papers no more than four pages long, was introduced explicitly to stop Cauchy filling it with his huge output. But when the rule was introduced, Cauchy just wrote lots of short papers. From his prolific pen the main outlines of complex analysis quickly emerged. And it is a simpler, more elegant and in many ways more complete theory than real analysis, where the whole idea started.

For instance, in real analysis a function may be differentiable, but its derivative may not be. It may be differentiable 23 times, but not 24. It may be differentiable as many times as you wish, but not possess a power series representation. None of these nasty things can happen in complex analysis. If a function is differentiable, then it can be differentiated as many times as you wish; moreover, it has a power series representation. The reason – closely related to Cauchy's Theorem and probably a fact used by Gauss in his unknown proof – is that in order to be differentiable, a complex function must satisfy some very stringent conditions, known as the *Cauchy–Riemann equations*. These equations lead directly to Gauss's result that the integral between two points may depend on the path chosen. Equivalently, as Cauchy noted, the integral round a *closed* path need not be zero. It is zero provided the function concerned is differentiable (so in particular is not infinite) at all points inside the path.

There was even a theorem – the residue theorem – that told you the value of an integral round a closed path, and it depended only on the locations of the points at which the function became infinite, and its behaviour near those points. In short, the entire structure of a complex function is determined by its singularities – the points at which it is badly behaved. And the most important singularities are its poles, the places where it becomes infinite.

The square root of minus one puzzled mathematicians for centuries. Although there seemed to be no such number, it kept turning up in calculations. And there were hints that the concept must make some kind of sense, because it could be used to obtain perfectly valid results which did not themselves involve taking the square root of a negative number.

As the successful uses of this impossible quantity continued to grow, mathematicians began to accept it as a useful device. Its status remained uncertain until it was realized that there is a logically consistent extension of the traditional system of real numbers, in which the square root of minus one is a new kind of quantity – but one that obeys all of the standard laws of arithmetic.

Geometrically, the real numbers form a line and the complex numbers form a plane; the real line is one of the two axes of this plane. Algebraically, complex numbers are just pairs of real numbers with particular formulas for adding the pairs or multiplying them.

Now accepted as sensible quantities, complex numbers quickly spread throughout mathematics because they simplified calculations by avoiding the need to consider positive and negative numbers separately. In this respect they can be considered as analogous to the earlier invention of negative numbers, which avoided the need to consider addition and subtraction separately. Today, complex numbers, and the calculus of complex functions, are routinely used as an indispensable technique in virtually all branches of science, engineering and mathematics.

What complex numbers do for us

Today, complex numbers are widely used in physics and engineering. A simple example occurs in the study of oscillations: motions that repeat periodically. Examples include the shaking of a building in an earthquake, vibrations in cars and the transmission of alternating electrical current.

The simplest and most fundamental type of oscillation takes the form $a \cos \omega t$, where t is time, a is the amplitude of the oscillation and ω is its frequency. It turns out to be convenient to rewrite this formula as the real part of the complex function $e^{i\omega t}$. The use of complex numbers simplifies calculations because the exponential function is simpler than the cosine. So engineers studying oscillations prefer to work with complex exponentials, and revert to the real part only at the end of the calculation.

Complex numbers also determine the stabilities of steady states of dynamical systems, and are widely used in control theory. This subject is about methods for stabilizing systems that would otherwise be unstable. An example is the use of computer-controlled moving control surfaces to stabilize the space shuttle in flight. Without this application of complex analysis, the space shuttle would fly like a brick.

CHAPTER 11

Firm Foundations

Making calculus make sense

By 1800 mathematicians and physicists had developed calculus into an indispensable tool for the study of the natural world, and the problems that arose from this connection led to a wealth of new concepts and methods – for example, ways to solve differential equations – that made calculus one of the richest and hottest research areas in the whole of mathematics. The beauty and power of calculus had become undeniable. However, Bishop Berkeley's criticisms of its logical basis remained unanswered, and as people began to tackle more sophisticated topics, the whole edifice started to look decidedly wobbly. The early cavalier use of infinite series, without regard to their meaning, produced nonsense as well as insights. The foundations of Fourier analysis were non-existent, and different mathematicians were claiming proofs of contradictory theorems. Words like 'infinitesimal' were bandied about without being defined; logical paradoxes abounded; even the meaning of the word 'function' was in dispute. Clearly these unsatisfactory circumstances could not go on indefinitely.

Sorting it all out took a clear head, and a willingness to replace intuition by precision, even if there was a cost in comprehensibility. The main players were Bernard Bolzano, Augustin-Louis Cauchy, Niels Abel, Peter Dirichlet and, above all, Karl Weierstrass. Thanks to their efforts, by 1900 even the most complicated manipulations of series, limits, derivatives and integrals could be carried out safely, accurately and without paradoxes. A new subject was created: analysis. Calculus became one core aspect of analysis, but more subtle and more basic concepts, such as continuity and limits, took logical precedence, underpinning the ideas of calculus. Infinitesimals were banned, completely.

Fourier

Before Fourier stuck his oar in, mathematicians were fairly happy that they knew what a function was. It was some kind of process, f, which took a number, x, and produced another number, $f(x)$. Which numbers, x, make sense depends on what f is. If $f(x) = 1/x$, for instance, then x has to be non-zero. If $f(x) = \sqrt{x}$, and we are working with real numbers, then x must be positive. But when pressed for a definition, mathematicians tended to be a little vague.

The source of their difficulties, we now realize, was that they were grappling with several different features of the function concept – not just what a rule associating a number x with another number $f(x)$ is, but what properties that rule possesses: continuity, differentiability, capable of being represented by some type of formula and so on. In particular, they were uncertain how to handle discontinuous functions, such as

$$f(x) = 0 \text{ if } x \leq 0, \quad f(x) = 1 \text{ if } x > 0$$

This function suddenly jumps from 0 to 1 as x passes through 0. There was a prevailing feeling that the obvious reason for the jump was the change in the formula: from $f(x) = 0$ to $f(x) = 1$. Alongside that was the feeling that this is the only way that jumps can appear; that any single formula automatically avoided such jumps, so that a small change in x always caused a small change in $f(x)$.

Another source of difficulty was complex functions, where – as we have seen – natural functions like the square root are two-valued, and logarithms are infinitely many valued. Clearly the logarithm must be a function – but when there are *infinitely many* values, what is *the rule* for getting $f(z)$ from z? There seemed to be infinitely many different rules, all equally valid. In order for these conceptual difficulties to be resolved, mathematicians had to have their noses firmly rubbed in them to experience just how messy the real situation was. And it was Fourier who really got up their noses, with his amazing ideas about writing *any* function as an infinite series of sines and cosines, developed in his study of heat flow.

Fourier's physical intuition told him that his method should be very general indeed. Experimentally, you can imagine holding the temperature of a metal bar at 0 degrees along half of its length, but 10 degrees, or 50, or whatever along the rest of its length. Physics did not seem to be bothered by discontinuous functions, whose formulas suddenly changed. Physics *did not work with formulas* anyway. *We* use formulas to model physical reality, but that's just technique, it's how we like to think. Of course the temperature will fuzz out a little at the junction of these two regions, but mathematical models are always approximations to physical reality. Fourier's method of trigonometric series, applied to a discontinuous function of this kind, seemed to give perfectly sensible results. Steel bars really did smooth out the temperature distribution the way his heat equation, solved using trigonometric series, specified. In *The Analytical Theory of Heat* he made his position plain: 'In general, the function $f(x)$ represents a succession of values or ordinates each of which is arbitrary. We do not suppose these ordinates to be subject to a common law. They succeed each other in any manner whatever'.

Bold words; unfortunately, his evidence in their support did not amount to a mathematical proof. It was, if anything, even more sloppy than the reasoning employed by people like Euler and Bernoulli. Additionally, if Fourier was right, then his series in effect derived a common law for discontinuous functions. The function above, with values 0 and 1, has a periodic relative, the *square wave*. And the square wave has a single Fourier series, quite a nice one, which works equally well in those regions where the function is 0 and in those regions where the function is 1. So a function that *appears* to be represented by two different laws can be rewritten in terms of one law.

Slowly the mathematicians of the 19th century started separating the different conceptual issues in this difficult area. One was the *meaning* of the term, function. Another was the various ways of *representing* a function – by a formula, a power series, a Fourier series or whatever. A third was what *properties* the function possessed. A fourth was which representations *guaranteed* which properties. A single polynomial, for example, defines a continuous function. A single Fourier series, it seemed, might not.

Fourier analysis rapidly became the test case for ideas about the function concept. Here the problems came into sharpest relief and esoteric technical distinctions turned out to be important. And it was in a paper on Fourier series, in 1837, that Dirichlet introduced the modern definition of a function. In effect, he agreed with Fourier:

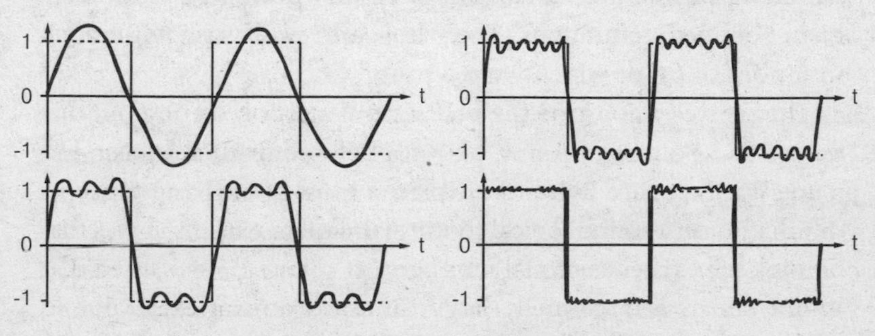

The square wave and some of its Fourier approximations

a variable y is a function of another variable x if for each value of x (in some particular range) there is specified a *unique* value of y. He explicitly stated that no particular law or formula was required: it is enough for y to be specified by some well-defined sequence of mathematical operations, applied to x. What at the time must have seemed an extreme example is 'one he made earlier', in 1829: a function $f(x)$ taking one value when x is rational, and a different value when x is irrational. This function is discontinuous at every point. (Nowadays functions like this are viewed as being rather mild; far worse behaviour is possible.)

For Dirichlet, the square root was not one two-valued function. It was two one-valued functions. For real x, it is natural – but not essential – to take the positive square root as one of them, and the negative square root as the other. For complex numbers, there are no obvious natural choices, although a certain amount can be done to make life easier.

Continuous functions

By now it was dawning on mathematicians that although they often stated definitions of the term 'function', they had a habit of assuming extra properties that did not follow from the definition. For example, they assumed that any sensible formula, such as a polynomial, automatically defined a continuous function. But they had never proved this. In fact, they couldn't prove it, because they hadn't defined 'continuous'. The whole area was awash with vague intuitions, most of which were wrong.

The person who made the first serious start on sorting out this mess was a Bohemian priest, philosopher and mathematician. His name was Bernhard Bolzano. He placed most of the basic concepts of calculus on a sound logical footing; the main exception was that he took the existence of real numbers for granted. He insisted that infinitesimals and infinitely large numbers do not exist, and so cannot be used, however evocative they may be. And he gave the first effective definition of a continuous function. Namely, f is continuous if the difference $f(x + a) - f(x)$ can be made as small as we please by choosing a sufficiently small. Previous authors had tended to say things like 'if a is infinitesimal then $f(x + a) - f(x)$ is infinitesimal'. But for Bolzano, a was just a number, like any others. His point was that whenever you specify how small you want $f(x + a) - f(x)$ to be, you must then specify a suitable value for a. It wasn't necessary for the *same* value to work in every instance.

So, for instance, $f(x) = 2x$ is continuous, because $2(x + a) - 2x = 2a$. If you want $2a$ to be smaller than some specific number, say 10^{-10}, then you have to make a smaller than $10^{-10}/2$. If you try a more complicated function, like $f(x) = x^2$, then the exact details are a bit complicated because the right a depends on x as well as on the chosen size, 10^{-10}, but any competent mathematician can work this out in a few minutes. Using this definition, Bolzano *proved* – for the first time ever – that a polynomial function is continuous. But for 50 years, no one noticed. Bolzano had published his work in a journal that mathematicians hardly ever read, or had access to. In these days of the Internet it is difficult to realize just how poor communications were even 50 years ago, let alone 180.

In 1821 Cauchy said much the same thing, but using slightly confusing terminology. His definition of continuity of a function f was that $f(x)$ and $f(x + a)$ differ by an infinitesimal amount whenever a is infinitesimal, which at first sight looks like the old, poorly-defined approach. But infinitesimal to Cauchy did not refer to a *single* number that was somehow infinitely small, but to an ever-decreasing sequence of numbers. For instance, the sequence 0.1, 0.01, 0.001, 0.0001 and so on is infinitesimal in Cauchy's sense, but each of the individual numbers, such as 0.0001, is just a conventional real number – small, perhaps, but not infinitely so. Taking this terminology into account, we see that Cauchy's concept of continuity amounts to exactly the same thing as Bolzano's.

Another critic of sloppy thinking about infinite processes was Abel, who complained that people were using infinite series without enquiring whether the sums made any sense. His criticisms struck home, and gradually order emerged from the chaos.

Limits

Bolzano's ideas set these improvements in motion. He made it possible to define the limit of an infinite sequence of numbers, and from that the series,which is the sum of an infinite sequence. In particular, his formalism implied that

$$1 + \frac{1}{2} + \frac{1}{4} + \frac{1}{8} + \frac{1}{16} + \ \ldots$$

carries on forever, is a meaningful sum, and its value is exactly 2. Not a little bit less; not an infinitesimal amount less; just plain 2.

To see how this goes, suppose we have a sequence of numbers

$$a_0, a_1, a_2, a_3, \ldots$$

going on forever. We say that a_n tends to *limit* a as n tends to infinity if, given any number $\varepsilon > 0$ there exists a number N such that the difference between a_n and a is less than ε whenever $n > N$. (The symbol ε, which is the one traditionally used, is the Greek epsilon.) Note that all of the numbers in this definition are finite – there are no infinitesimals or infinities.

To add up the infinite series above, we look at the finite sums

$$a_0 = 1$$

$$a_1 = 1 + \frac{1}{2} = \frac{3}{2}$$

$$a_2 = 1 + \frac{1}{2} + \frac{1}{4} = \frac{7}{4}$$

$$a_3 = 1 + \frac{1}{2} + \frac{1}{4} + \frac{1}{8} = \frac{15}{8}$$

What analysis did for them

The mathematical physics of the 19th century led to the discovery of a number of important differential equations. In the absence of high-speed computers, able to find numerical solutions, the mathematicians of the time invented new special functions to solve these equations. These functions are still in use today. An example is *Bessel's equation*, first derived by Daniel Bernoulli and generalized by Bessel. This takes the form

$$x^2 \frac{d^2y}{dx^2} + x\frac{dy}{dx} + (x^2 - k^2)y = 0$$

and the standard functions, such as exponentials, sines, cosines and logarithms, do not provide a solution.

However, it is possible to use analysis to find solutions, in the form of power series. The power series determine new functions, *Bessel functions*. The simplest types of Bessel function are denoted $J_k(x)$; there are several others. The power series permits the calculation of $J_k(x)$ to any desired accuracy.

Bessel functions turn up naturally in many problems about circles and cylinders, such as the vibration of a circular drum, propagation of electromagnetic waves in a cylindrical waveguide, the conduction of heat in a cylindrical metal bar, and the physics of lasers.

and so on. The difference between a_n and 2 is $1/2^n$. To make this less than ε we take $n > N = \log_2(1/\varepsilon)$.

A series that has a finite limit is said to be *convergent*. A finite sum is defined to be the limit of the sequence of finite sums, obtained by adding more and more terms. If that limit exists, the series is convergent. Derivatives and integrals are just limits of various kinds. They exist – that is, they make mathematical sense – provided those finite sums converge. Limits, just as Newton maintained, are about what certain quantities *approach* as some other number approaches infinity, or zero. The number does not have to *reach* infinity or zero.

The whole of calculus now rested on sound foundations. The downside was that whenever you used a limiting process, you had to make sure that it converged. The best way to do that was to prove ever more general theorems about what kinds of functions are continuous, or differentiable or integrable, and which sequences or series converge. This is what the analysts proceeded to do, and it is why we no longer worry about the difficulties pointed out by Bishop Berkeley. It is also why we no longer argue about Fourier series: we have a sound idea of when they converge, when they do not and indeed *in what sense* they converge. There are several variations on the basic theme, and for Fourier series you have to pick the right ones.

Power series

Weierstrass realized that the same ideas work for complex numbers as well as real numbers. Every complex number $z = x + iy$ has an absolute value $|z| = \sqrt{x^2 + y^2}$, which by Pythagoras's Theorem is the distance from 0 to z in the complex plane. If you measure the size of a complex expression using its absolute value, then the real-number concepts of limits, series and so on, as formulated by Bolzano, can immediately be transferred to complex analysis.

Weierstrass noticed that one particular type of infinite series seemed to be especially useful. It is known as a power series, and it looks like a polynomial of infinite degree:

$$f(z) = a_0 + a_1 z + a_2 z^2 + a_3 z^3 + \cdots$$

where the coefficients a_n are specific numbers. Weierstrass embarked on a huge programme of research, aiming to found the whole of complex analysis on power series. It worked brilliantly.

For example, you can define the exponential function by

$$e^z = 1 + z + \frac{1}{2} z^2 + \frac{1}{6} z^3 + \frac{1}{24} z^4 + \frac{1}{120} z^5 + \cdots$$

where the numbers 2, 4, 6 and so on are *factorials*: products of consecutive integers (for instance $120 = 1\times2\times3\times4\times5$). Euler had already obtained this formula heuristically; now Weierstrass could make rigorous sense of it. Again taking a leaf from Euler's book, he could then relate the trigonometric functions to the exponential function, by defining

$$\cos\theta = \frac{1}{2}(e^{i\theta} + e^{-i\theta})$$

$$\sin\theta = \frac{1}{2i}(e^{i\theta} - e^{-i\theta})$$

All of the standard properties of these functions followed from their power series expression. You could even define π, and prove that $e^{i\pi} = -1$, as Euler had maintained. And that in turn meant that complex logarithms did what Euler had claimed. *It all made sense.* Complex analysis was not just some mystical extension of real analysis: it was a sensible subject in its own right. In fact, it was often simpler to work in the complex domain, and read off the real result at the end.

To Weierstrass, all this was just a beginning – the first phase of a vast programme. But what mattered was getting the foundations right. If you did that, the more sophisticated material would readily follow.

Weierstrass was unusually clear-headed, and he could see his way through complicated combinations of limits and derivatives and integrals without getting confused. He could also spot potential difficulties. One of his most surprising theorems proves that there exists a function $f(x)$ of a real variable x, which is continuous at

The Riemann Hypothesis

The most famous unsolved problem in the whole of mathematics is the Riemann Hypothesis, a problem in complex analysis which arose in connection with prime numbers but has repercussions throughout mathematics.

Around 1793 Gauss conjectured that the number of primes less than x is approximately $x/\log x$. In fact he suggested a more accurate approximation called the logarithmic integral. In 1737 Euler had noticed an intriguing connection between number theory and analysis: the infinite series

$$1 + 2^{-s} + 3^{-s} + 4^{-s} + \cdots$$

is equal to the product, over all primes p, of the series

$$1 + p^{-s} + p^{-2s} + p^{-3s} + \ldots = \frac{1}{1 - p^{-s}}.$$

Here we must take $s > 1$ for the series to converge.

In 1848 Pafnuty Chebyshev made some progress towards a proof of Gauss's conjecture, using a complex function related to Euler's series, later called the *zeta function* $\zeta(z)$. The role of this function was made clear by Riemann in his 1859 paper *On the Number of Primes Less Than a Given Magnitude*. He showed that the statistical properties of primes are closely related to the zeros of the zeta function, that is, the solutions, z, of the equation $\zeta(z) = 0$.

In 1896 Jacques Hadamard and Charles de la Vallée Poussin used the zeta function to prove the Prime Number Theorem. The main step is to show that $\zeta(z)$ is non-zero for all z of the form $1 + it$. The more control we can gain over the location of the zeros of the zeta function, the more we learn about primes. Riemann conjectured that all zeros, other than some obvious ones at negative even integers, lie on the critical line $z = {}^1/_2 \pm it$.

In 1914 Hardy proved that an infinite number of zeros lie on this line. Extensive computer evidence also supports the conjecture. Between 2001 and 2005 Sebastian Wedeniwski's program ZetaGrid verified that the first 100 billion zeros lie on the critical line.

The Riemann Hypothesis was part of Problem 8 in Hilbert's famous list of 23 great unsolved mathematical problems, and is one of the Millennium prize problems of the Clay Mathematics Institute.

every point, but differentiable at no point. The graph of f is a single, unbroken curve, but it is such a wiggly curve that it has no well-defined tangent anywhere. His predecessors would not have believed it; his contemporaries wondered what it was for. His successors developed it into one of the most exciting new theories of the 20th century, fractals.

But more of that story later.

A firm basis

The early inventors of calculus had taken a rather cavalier approach to infinite operations. Euler had assumed that power series were just like polynomials, and used this assumption to devastating effect. But in the hands of ordinary mortals, this kind of assumption can easily lead to nonsense. Even Euler stated some pretty stupid things. For instance, he started from the power series

$$1 + x + x^2 + x^3 + x^4 + \cdots$$

which sums to $1/(1 - x)$, set $x = -1$ and deduced that

$$1 - 1 + 1 - 1 + 1 - 1 + \cdots = \tfrac{1}{2}$$

which is nonsense. The power series does not converge unless x is strictly between -1 and 1, as Weierstrass's theory makes clear.

What analysis does for us

Analysis is used in biology to study the growth of populations of organisms. A simple example is the logistic or Verhulst–Pearl model. Here the change in the population, x, as a function of time t is modelled by a differential equation

$$\frac{dx}{dt} = kx\left(1 - \frac{x}{M}\right)$$

where the constant M is the carrying capacity, the largest population that the environment can sustain.

Standard methods in analysis produce the explicit solution

$$x(t) = \frac{Mx_0}{x_0 + (M - x_0)e^{-kt}}$$

which is called the logistic curve. The corresponding pattern of growth begins with rapid (exponential) growth, but when the population reaches half the carrying capacity it begins to level off, and eventually levels off at the carrying capacity.

This curve is not totally realistic, although it fits many real populations quite well. More complicated models of the same kind provide closer fits to real data. Human consumption of natural resources can also follow a pattern similar to the logistic curve, making it possible to estimate future demand, and how long the resources are likely to last.

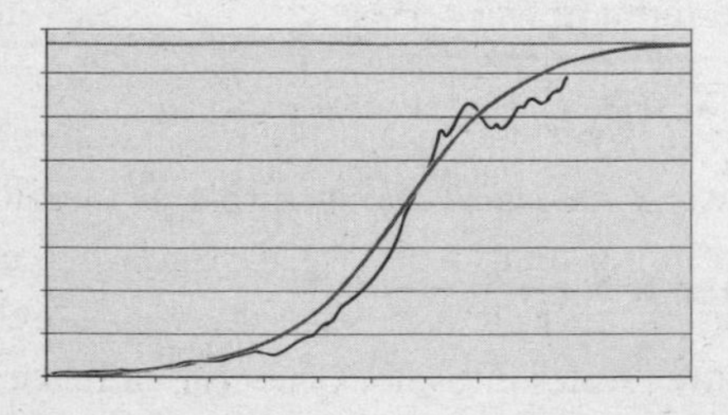

World consumption of crude oil, 1900–2000:smooth curve, logistic equation; jagged curve, actual data

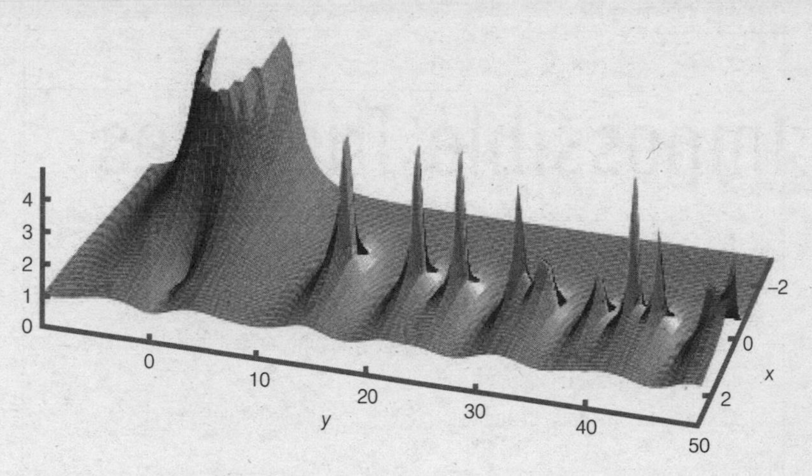

Absolute value of the Riemann zeta function

The zeta landscape - graph of $|1/\zeta(x+iy)|$. Peaks correspond to zeros of the zeta function

Taking criticisms like those made by Bishop Berkeley seriously did, in the long run, enrich mathematics and place it on a firm footing. The more complicated the theories became, the more important it was to make sure you were standing on firm ground.

Today, most users of mathematics once more ignore such subtleties, safe in the knowledge that they have been sorted and that anything that looks sensible probably has a rigorous justification. They have Bolzano, Cauchy and Weierstrass to thank for this confidence. Meanwhile, professional mathematicians continue to develop rigorous concepts about infinite processes. There is even a movement to revive the concept of an infinitesimal, known as non-standard analysis, and it is perfectly rigorous and technically useful on some otherwise intractable problems. It avoids logical contradictions by making infinitesimals a new kind of number, not a conventional real number. In spirit, it is close to the way Cauchy thought. It remains a minority speciality – but watch this space.

CHAPTER 12

Impossible Triangles

Is Euclid's geometry the only one?

Calculus was based on geometric principles, but the geometry was reduced to symbolic calculations, which were then formalized as analysis. However, the role of visual thinking in mathematics was also developing, in a new and initially rather shocking direction. For more than 2000 years the name Euclid had been synonymous with geometry. His successors developed his ideas, especially in their work on conic sections, but they did not make radical changes to the concept of geometry itself. Essentially, it was assumed that there can be only one geometry, Euclid's, and that this is an exact mathematical description of the true geometry of physical space. People found it difficult even to conceive of alternatives.

It couldn't last.

Spherical and projective geometry

The first significant departure from Euclidean geometry arose from the very practical issue of navigation. Over short distances, the Earth is almost flat, and its geographical features can be mapped on a plane. But as ships made ever longer voyages, the true shape of the planet had to be taken into account. Several ancient civilizations knew that the Earth is round – there is ample evidence, from the way ships seem to disappear over the horizon to the shadow of the planet on the Moon during lunar eclipses. It was generally assumed that the Earth is a perfect sphere.

In reality, the sphere is slightly flattened: the diameter at the equator is 12,756 km, whereas that at the poles is 12,714 km. The difference is relatively small – one part in 300. In times when navigators routinely made errors of several hundred kilometres, a spherical Earth provided an entirely acceptable mathematical model. At that time, however, the emphasis was on spherical *trigonometry* rather than geometry – the nuts and bolts of navigational calculations, not the logical analysis of the sphere as a kind of space. Because the sphere sits naturally within three-dimensional Euclidean space, no one considered spherical geometry to be different from Euclidean. Any differences were the result of the Earth's curvature. The geometry of space *itself* remained Euclidean.

A more significant departure from Euclid was the introduction, from the early 17th century onwards, of *projective geometry*. This topic emerged not from science but from art: theoretical and practical investigations of perspective by the Renaissance artists of Italy. The aim was to make paintings look realistic; the outcome was a new way to think about geometry. But, again, this development could be seen as an innovation within the classical Euclidean frame. It was about how we *view* space, not about space.

The discovery that Euclid was not alone, that there can exist logically consistent types of geometry in which many of Euclid's theorems fail to hold, emerged from a renewed interest in the logical foundations of geometry, debated and developed from the middle of the 18th century to the middle of the 19th. The big issue was Euclid's Fifth Postulate, which – clumsily – asserted the existence of parallel lines. Attempts to deduce the Fifth Postulate from the remainder of Euclid's axioms eventually led to the realization that no such deduction is possible. There are consistent types of geometry other than Euclidean. Today, these non-Euclidean geometries have become indispensable tools in pure mathematics and mathematical physics.

Geometry and art

As far as Europe was concerned, geometry was becalmed in the doldrums between the years 300 and 1600. The revival of geometry as a living subject came from the question of perspective in art: how to render a three-dimensional world realistically on two-dimensional canvas.

The artists of the Renaissance did not just create paintings. Many were employed to carry out engineering works, for peaceful or warlike purposes. Their art had a practical side, and the geometry of perspective was a practical quest, applying to architecture as well as to the visual arts. There was also a growing interest in optics, the mathematics of light, which blossomed once the telescope and microscope were invented. The first major artist to think about the mathematics of perspective was Filippo Brunelleschi. In fact, his art was mostly a vehicle for his mathematics. A seminal book is Leone Battista Alberti's *Della Pittura*, written in 1435 and printed in 1511. Alberti began by making some important, and relatively harmless, simplifications – the standard reflex of a true mathematician. Human vision is a complex subject. For example, we use two slightly separated eyes to generate stereoscopic images, providing a feeling of depth. Alberti simplified reality by assuming a single eye with a pinprick pupil, which worked like a pinhole camera. He imagined an artist painting a scene, setting up his easel and trying to make the image on the canvas match the one perceived by his (single) eye. Both canvas and reality project their images on to the retina, at the back of the eye. The simplest (conceptual) way to ensure a perfect match is to make the canvas transparent, look through it from a fixed location and draw upon the canvas exactly what the eye sees. So the three-dimensional scene is *projected* on to the canvas. Join each feature of the scene to the eye by a straight line, note where this line meets the plane of the canvas: that's where you paint that feature.

This idea is not terribly practical if you take it literally, although some artists did just that, using translucent materials, or glass, in place of a canvas. They often did this as a preliminary step, transferring the resulting outline to canvas for proper painting. A more practical approach is to use this conceptual formulation to relate the geometry of the three-dimensional scene to that of the two-dimensional image. Ordinary Euclidean geometry is about features that remain unchanged by rigid motions – lengths, angles. Euclid did not formulate it that way, but his use of congruent triangles as a basic tool has the same effect. (These are triangles of the same size and shape, but in different locations.) Similarly, the geometry of perspective boils down to features that remain unchanged by projection. It is easy to see that lengths and angles do not behave like that. You can cover the Moon with your thumb, so lengths can change. Angles fare no better – when you look at the corner of a building, a right angle, it only *looks* like a right angle if you view it square on.

What properties of geometrical figures, then, are preserved by projection? The most important ones are so simple that it is easy to miss their significance. Points remain points. Straight lines remain straight. The image of a point lying on a straight lines lies on the image of that line. Therefore, if two lines meet at a point, their images meet at the corresponding point. Incidence relations of points and lines are preserved by projection.

An important feature that is not *quite* preserved is the relation 'parallel'. Imagine standing in the middle of a long, straight road and look ahead. The two sides of the road, which in three-dimensional reality are parallel – so never meet – do not appear parallel. Instead, they converge towards a single point on the distant horizon. They behave like this on an ideal infinite plane, not just on a slightly rounded Earth. In fact, they only behave exactly like this on a plane. On a sphere, there would be a tiny gap, too small to see, where the lines cross the horizon. And the whole issue of parallel lines on a sphere is tricky anyway.

This feature of parallel lines is very useful in perspective drawing. It lies behind the usual way of drawing right-angled boxes in perspective, using a horizon line and two vanishing points, which are where parallel edges of the box cross the horizon in perspective. Piero della Francesca's *De Prospettiva Pingendi* (1482–87) developed Alberti's methods into practical techniques for artists, and he used them to great effect in his dramatic and very realistic paintings.

The writings of the Renaissance painters solved many problems in the geometry of perspective, but they were semi-empirical, lacking the kind of logical foundation that Euclid had supplied for ordinary geometry. These foundational issues were finally resolved by Brook Taylor and Johann Heinrich Lambert in the 18th century. But by then, more exciting things were going on in geometry.

Desargues

The first non-trivial theorem in projective geometry was found by the engineer/architect Girard Desargues and published in 1648 in a book by Abraham Bosse. Desargues proved the following remarkable theorem. Suppose that triangles ABC and $A'B'C'$ are in perspective, which means that the three lines AA', BB' and CC' all pass through the same point O. Then the three points P, Q and R at which corresponding sides of the two triangles meet all lie on the same line. This result is called Desargues's Theorem to this day. It mentions no lengths, no angles – it is purely about incidence relations among lines and points. So it is a projective theorem.

There is a trick which makes the theorem obvious: imagine it as a drawing of a three-dimensional figure, in which the two triangles lie in two planes. Then the line along which those planes intersect is the line containing Desargues's three points P, Q and R. With a little care,

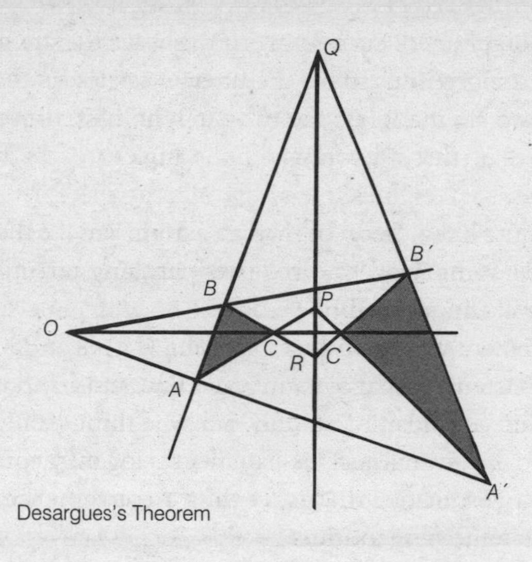

Desargues's Theorem

the theorem can even be proved this way, by constructing a suitable three-dimensional figure whose projection looks like the two triangles. So we can use Euclidean methods to prove projective theorems.

Euclid's axioms

Projective geometry differs from Euclidean geometry as far as its viewpoint goes (pun intended), but it is still related to Euclidean geometry. It is the study of new kinds of transformation, projections, but the underlying model of the space that is being transformed is Euclidean. Nevertheless, projective geometry made mathematicians more receptive to the possibility of new kinds of geometric thinking. And an old question, one that had lain dormant for centuries, once more came to the fore.

Nearly all of Euclid's axioms for geometry were so obvious that no sane person could seriously question them. All right angles are equal, for instance. If that axiom failed, there had to be something wrong with the definition of a right angle. But the Fifth Postulate, the one that was really about parallel lines, had a distinctly different flavour. It was complicated. Euclid states it this way: If a straight line falling on two straight lines make the interior angles on the same side less than two right angles, the two straight lines, if produced indefinitely, meet on that side on which the angles are less than the two right angles.

It sounded more like a theorem than an axiom. *Was* it a theorem? Might there be some way to prove it, perhaps starting from something simpler, more intuitive?

One improvement was introduced by John Playfair in 1795. He substituted the statement that for any given line, and any point not on that line, there exists one and only one line through the point that is parallel to the given line. This statement is logically equivalent to Euclid's Fifth Postulate – that is, each is a consequence of the other, given the remaining axioms.

Legendre

In 1794 Adrien-Marie Legendre discovered another equivalent statement, the existence of *similar triangles* – triangles having the same angles, but with edges of different sizes. But he, and most other mathematicians, wanted something even more intuitive. In fact, there was a feeling that the Fifth Postulate was simply superfluous – a consequence of the other axioms. All that was missing was a proof. So Legendre tried all sorts of things. Using only the other axioms, he proved – to his own satisfaction, at any rate – that the angles of a triangle either add up to 180° or less. (He must have known that in spherical geometry the sum is greater, but that is the geometry of the sphere, not the plane.) If the sum is always 180°, the Fifth Postulate follows. So he assumed that the sum could be less than 180°, and developed the implications of that assumption.

A striking consequence was a relation between the triangle's area and the sum of its angles. Specifically, the area is proportional to the amount by which the angle sum falls short of 180°. This seemed promising: if he could construct a triangle whose sides were twice those of a given triangle, but with the same angles, then he would obtain a contradiction, because the larger triangle would not have the same area as the smaller one. But however he tried to construct the larger triangle, he found himself appealing to the Fifth Postulate.

He did manage to salvage one positive result from the work. Without assuming the Fifth Postulate, he proved that it is impossible for some triangles to have angle-sums greater than 180°, while others have angle-sums less than 180°. If one triangle has angles that summed to more than 180°, so does every triangle; similarly if the sum is less than 180°. So there are three possible cases:

- The angles of every triangle add up to 180° exactly (Euclidean geometry)
- The angles of every triangle add up to less than 180°
- The angles of every triangle add up to more than 180° (a case that Legendre thought he had excluded; it later turned out that he had made other unstated assumptions to do so).

Saccheri

In 1733 Gerolamo Saccheri, a Jesuit priest at Pavia, published a heroic effort, *Euclides ab Omni Naevo Vindicatus* (*Euclid Vindicated from All Flaws*). He also considered three cases, of which the first was Euclidean geometry, but he used a quadrilateral to make the

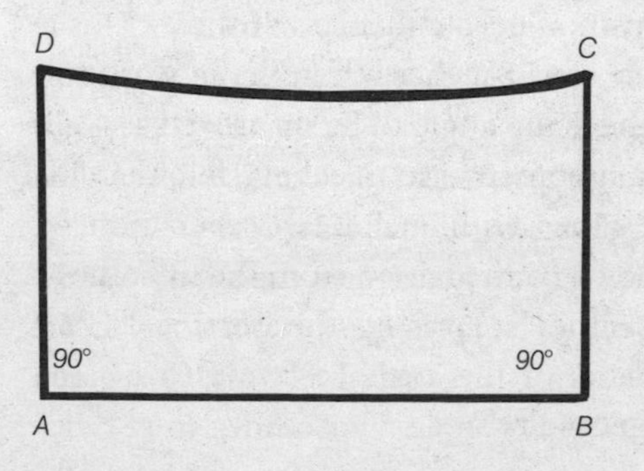

Saccheri's quadrilateral: the line CD has been drawn curved to avoid Euclidean assumptions about angles C and D

distinction. Suppose the quadrilateral is *ABCD*, with *A* and B right angles and *AC* = *BD*. Then, said Saccheri, Euclidean geometry implies that angles C and D are right angles. Less obviously, if C and D are right angles in any one quadrilateral of this kind, then the Fifth Postulate follows.

Without using the Fifth Postulate, Saccheri proved that angles C and D are equal. So that left two distinct possibilities:

- *Hypothesis of the obtuse angle*: both C and D are greater than a right angle.
- *Hypothesis of the acute angle*: both C and D are less than a right angle.

Saccheri's idea was to assume each of these hypotheses in turn, and deduce a logical contradiction. That would then leave Euclidean geometry as the only logical possibility.

He began with the hypothesis of the obtuse angle, and in a series of theorems deduced – so he thought – that angles C and D must in fact be right angles after all. This was a contradiction, so the hypothesis of the obtuse angle had to be false. Next, he assumed the hypothesis of the acute angle, which led to another series of theorems, all correct, and fairly interesting in their own right. Eventually he proved a rather complicated theorem about a family of lines all passing through one point, which implied that two of these lines would have a common perpendicular at infinity. This is not actually a contradiction, but Saccheri thought it was, and declared the hypothesis of the acute angle to be disproved as well.

That left only Euclidean geometry, so Saccheri felt that his programme was vindicated, along with Euclid. But others noticed that he had not really obtained a contradiction from the hypothesis of the acute angle; just a rather surprising theorem. By 1759 d'Alembert declared the status of the Fifth Postulate to be 'the scandal of the elements of geometry'.

What Non-Euclidean geometry did for them

By 1813 Gauss was becoming ever more convinced that what he first called anti-Euclidean, then astral and finally non-Euclidean geometry was a logical possibility. He began to wonder what the true geometry of space was, and measured the angles of a triangle formed by three mountains near Göttingen – the Brocken, the Hohehagen and the Inselberg. He used line-of-sight measurement so the curvature of the Earth did not come into play. The sum of the angles that he measured was 15 seconds of arc greater than 180°. If anything, this was the obtuse-angle case, but the likelihood of observational errors made the whole exercise moot. Gauss needed a much bigger triangle, and much more accurate instruments to measure its angles.

Lambert

A German mathematician, Georg Klügel, read Saccheri's book, and offered the unorthodox and rather shocking opinion that belief in the truth of the Fifth Postulate was a matter of experience rather than logic. Basically, he was saying that something in the way we think about space makes us believe in the existence of parallel lines of the kind envisaged by Euclid.

In 1766 Johann Henrich Lambert, following up on Klügel's suggestion, embarked on an investigation that was similar to Saccheri's, but he started from a quadrilateral with three right angles. The remaining angle must either be a right angle (Euclidean geometry), acute or obtuse. Like Saccheri, he thought that the obtuse angle case led to a contradiction. More precisely, he decided that it led to spherical geometry, where it had long been known that the angles of a quadrilateral add up to more than 360°, because the angles of a triangle add up to more than 180°. Since the sphere is not the plane, the obtuse case is ruled out.

However, he did not claim the same for the acute angle case. Instead, he proved some curious theorems, the most striking being a formula for the area of a polygon with *n* sides. Add all the angles, and subtract this from 2n – 4 right angles: the result is proportional to the polygon's area. This formula reminded Lambert of a similar formula for spherical geometry: add all the angles, and subtract 2n – 4 right angles from this: again the result is proportional to the polygon's area. The difference is minor: the subtraction is performed in the opposite order. He was led to a remarkably prescient but obscure prediction: the geometry of the acute-angle case is the same as that on a sphere with *imaginary radius*.

He then wrote a short article about trigonometric functions of imaginary angles, obtaining some beautiful and perfectly consistent formulas. We now recognize these functions as the so-called hyperbolic functions, which can be defined without using imaginary numbers, and they satisfy all of Lambert's formulas. Clearly something interesting must lie behind his curious, enigmatic suggestion. But what?

Gauss's dilemma

By now the best-informed geometers were getting a definite feeling that Euclid's Fifth Postulate could not be proved from the remaining axioms. The acute-angle case seemed too self-consistent ever to lead to a contradiction. On the other hand, a sphere of imaginary radius was not the sort of object that could be proposed to justify that belief.

One such geometer was Gauss, who convinced himself from an early age that a logically consistent non-Euclidean geometry was possible, and proved numerous theorems in such a geometry. But, as he made clear in an 1829 letter to Bessel, he had no intention of publishing any of this work, because he feared what he called the 'clamour of the Boeotians'. Unimaginative people would not understand, and in their ignorance and hidebound adherence to tradition, they would ridicule the work. In this he may have been influenced by the overarching status of Kant's widely acclaimed work in philosophy; Kant had argued that the geometry of space must be Euclidean.

By 1799 Gauss was writing to the Hungarian Wolfgang Bolyai, telling him that the research 'seems rather to compel me to doubt the truth of geometry itself. It is true that I have come upon much which by most people would be held to constitute a proof [of the Fifth Postulate from the other axioms]; but in my eyes it is as good as nothing.'

Other mathematicians were less circumspect. In 1826 Nikolai Ivanovich Lobachevsky, at the University of Kazan in Russia, gave lectures on non-Euclidean geometry. He knew nothing of Gauss's work, but had proved similar theorems using his own methods. Two papers on the topic appeared in 1829 and 1835. Rather than starting riots, as Gauss had feared, these papers sank pretty much without trace. By 1840 Lobachevsky was publishing a book on the topic, in which he complained about the lack of interest. In 1855 he published a further book on the subject.

Independently, Wolfgang Bolyai's son János, an officer in the army, came up with similar ideas around 1825, writing them up in a 26-page paper that was published as an appendix to his father's geometry text *Tentamen Juventum Studiosam in Elementa Matheseos* (*Essay on the Elements of Mathematics for Studious Youths*) of 1832. 'I have made such wonderful discoveries that I am myself lost in astonishment,' he wrote to his father.

Gauss read the work, but explained to Wolfgang that it was impossible for him to praise the young man's efforts, because Gauss would in effect be praising himself. This was perhaps a little unfair, but it was how Gauss tended to operate.

Non-Euclidean geometry

The history of non-Euclidean geometry is too complicated to describe in any further detail, but we can summarize what followed these pioneering efforts. There is a deep unity behind the three cases noticed by Saccheri, by Lambert, and by Gauss, Bolyai and Lobachevsky. What unites them is the concept of *curvature*. Non-Euclidean geometry is really the natural geometry of a curved surface.

If the surface is positively curved like a sphere, then we have the case of the obtuse angle. This was rejected because spherical geometry differs from Euclidean in obvious ways – for example, any two lines, that is, great circles (circles whose centres are at the centre of the sphere), meet in two points, not the one that we expect of Euclidean straight lines.

Actually, we now realize that this objection is unfounded. If we identify diametrically opposite points of the sphere – that is, pretend they are identical – then lines (great circles) still make sense, because if a point lies on a great circle, so does the diametrically opposite point. With this identification, nearly all of the geometric properties remain unchanged, but now lines meet in *one* point. Topologically, the surface that results is the projective plane, although the geometry concerned is not orthodox projective

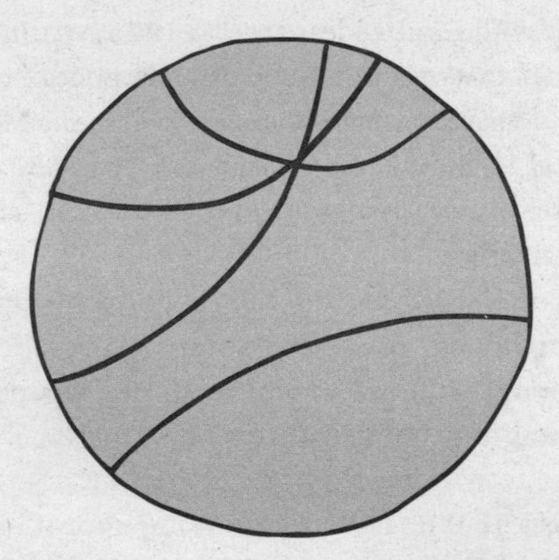

Poincaré's model of hyperbolic geometry makes it clear there are infinitely many parallel lines through a point that do not meet a given line

geometry. We now call it elliptic geometry, and it is considered just as sensible as Euclidean geometry.

If the surface is negatively curved, shaped like a saddle, then we have the case of the acute angle. The resulting geometry is called *hyperbolic*. It has numerous intriguing features, which distinguish it from Euclidean geometry.

If the surface has zero curvature, like a Euclidean plane, then it is the Euclidean plane, and we get Euclidean geometry.

All three geometries satisfy all of Euclid's axioms other than the Fifth Postulate. Euclid's decision to include his postulate is vindicated.

These various geometries can be modelled in several different ways. Hyperbolic geometry is especially versatile in this respect. In one model the space concerned is the upper half of the complex plane, omitting the real axis and everything below it. A line is a semicircle that meets the real axis at right angles. Topologically, this space is the same as a plane and the lines are identical to ordinary lines. The curvature of the lines reflects the negative curvature of the underlying space.

In a second model of hyperbolic geometry, introduced by Poincaré, the space is represented as the interior of a circle, not including its boundary, and the lines are circles that meet the boundary at right angles. Again, the distorted geometry reflects the curvature of the underlying space. The artist Maurits Escher produced many pictures based on this model of hyperbolic geometry, which he learned from the Canadian geometer Coxeter.

These two models hint at some deep connections between hyperbolic geometry and complex analysis. These connections relate to certain groups of transformations of the complex plane; hyperbolic geometry is the geometry of their invariants, according to Felix Klein's Erlangen Programme. Another class of transformations, called Möbius transformations, brings elliptic geometry into play as well.

What Non-Euclidean geometry does for us

What shape is the universe? The question may seem simple but answering it is difficult – partly because the universe is so big, but mainly because we are inside it and cannot stand back and see it as a whole. In an analogy that goes back to Gauss, an ant living on a surface, and observing it only within that surface, could not easily tell if the surface was a plane, a sphere, a torus or something more complicated.

General relativity tells us that near a material body, such as a star, space–time is curved. Einstein's equations, which relate the curvature to the density of matter, have many different solutions. In the simplest ones,

Space with positive, negative and zero curvature

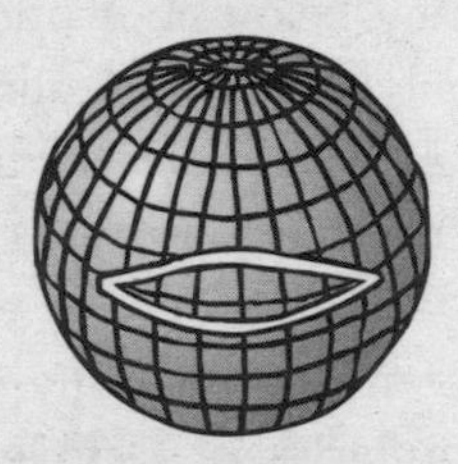

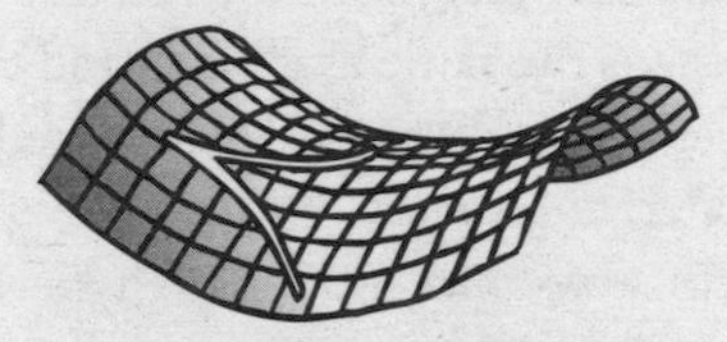

A *closed universe* curves back on itself. Lines that were diverging apart come back together. Density > critical density.

An *open universe* curves away from itself. Diverging lines curve at increasing angles away from each other. Density < critical density.

A *flat universe* has no curvature. Diverging lines remain at a constant angle with respect to each other. Density = critical density.

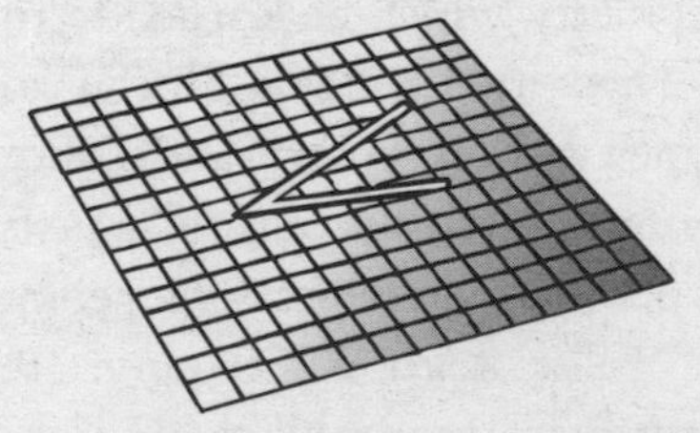

the universe as a whole has positive curvature and its topology is that of a sphere. But for all we can tell, the overall curvature of the real universe might be negative instead. We don't even know whether the universe is infinite, like Euclidean space, or is of finite extent, like a sphere. A few physicists maintain that the universe is infinite, but the experimental basis for this assertion is highly questionable. Most think it is finite.

To get Poincaré's dodecahedral space, identify opposite faces

Surprisingly, a finite universe can exist without having a boundary. The sphere is like that in two dimensions, and so is a torus. The torus can be given a *flat* geometry, inherited from a square by identifying opposite edges. Topologists have also discovered that space can be finite yet negatively curved: one way to construct such spaces is to take a finite polyhedron in hyperbolic space and identify various faces, so that a line passing out of the polyhedron across one face immediately re-enters at another face. This construction is analogous to the way the edges of the screen wrap round in many computer games.

If space is finite, then it should be possible to observe the same star in different directions, though it might seem much further away in some directions than in others, and the observable region of the universe might be too small anyway. If a finite space has hyperbolic geometry, these multiple occurrences of the same stars in different directions determine a system of gigantic circles in the heavens, and the geometry of those circles determines which hyperbolic space is being observed. But the circles could be anywhere among the billions of stars that can be seen, and so far attempts to observe them, based on statistical correlations among the apparent positions of stars, have not produced any result.

In 2003 data from the Wilkinson Microwave Anisotropy Probe led Jean-Pierre Luminet and his collaborators to propose that space is finite but *positively* curved. They found that Poincaré's dodecahedral space – obtained by identifying opposite faces of a curved dodecahedron – gives the best agreement with observations. This suggestion received wide publicity as the assertion that the universe is shaped like a football. This suggestion has not been confirmed and we currently have no idea of the true shape of space. However, we do have a much better understanding of what has to be done to find out.

The geometry of space

What of the geometry of space? We now agree with Klügel, and disdain Kant. This is a matter for experience, not something that can be deduced by thought alone. Einstein's General Relativity tells us that space (and time) can be curved; the curvature is the gravitational effect of matter. The curvature can vary from one location to another, depending on how the matter is distributed. So the geometry of space is not really the issue. Space can have different geometries in different places. Euclid's geometry works well on human scales, in the human world, because gravitational curvature is so small that we don't observe it in our daily lives. But out there in the greater universe, non-Euclidean geometries prevail.

To the ancients and indeed well into the 19th century, mathematics and the real world were hopelessly confused. There was a general belief that mathematics was a representation of basic and inevitable features of the real world, and that mathematical truth was absolute. Nowhere was this assumption more deeply rooted than in classical geometry. Space was Euclidean, to virtually everyone who thought about the question. What else could it be?

This question ceased to be rhetorical when logically consistent alternatives to Euclid's geometry began to appear. It took time to recognize that they *were* logically consistent – at least, just as consistent as Euclid's geometry – and even longer to realize that our own physical space might not be perfectly Euclidean. As always, human parochialism was to blame – we were projecting our own limited experiences in one tiny corner of the universe on to the universe as a whole. Our imaginations do seem to be biased in favour of a Euclidean model, probably because, on the small scales of our experience, it is an excellent model and also the simplest one available.

Thanks to some imaginative and unorthodox thinking, often viciously contested by a less imaginative majority, it is now understood – by mathematicians and physicists, at least – that there are many alternatives to Euclid's geometry, and that the nature of physical space is a question for observation, not thought alone. Nowadays, we make a clear distinction between mathematical models of reality, and reality itself. For that matter, much of mathematics bears no obvious relation to reality at all – but is useful, all the same.

CHAPTER 13

The Rise of Symmetry

How not to solve an equation

Around 1850 mathematics underwent one of the most significant changes in its entire history, although this was not apparent at the time. Before 1800, the main objects of mathematical study were relatively concrete: numbers, triangles, spheres. Algebra used formulas to represent manipulations with numbers, but the formulas themselves were viewed as symbolic representations of *processes*, not as things in their own right. But by 1900 formulas and transformations were viewed as *things*, not processes, and the objects of algebra were much more abstract and far more general. In fact, pretty much anything went as far as algebra was concerned. Even the basic laws, such as the commutative law of multiplication, $ab = ba$, had been dispensed with in some important areas.

Group Theory

These changes came about largely because mathematicians discovered group theory, a branch of algebra that emerged from unsuccessful attempts to solve algebraic equations, especially the quintic, or fifth degree, equation. But within 50 years of its discovery, group theory had been recognized as the correct framework for studying the concept of *symmetry*. As the new methods sank into the collective consciousness, it became clear that symmetry is a deep and central idea, with innumerable applications to the physical sciences, indeed to the biological ones as well. Today, group theory has become an indispensable tool in every area of mathematics and science, and its connections with symmetry are emphasized in most introductory texts. But this point of view took several decades to develop. Around 1900 Henri Poincaré said that group theory was effectively the whole of mathematics reduced to its essentials, which was a bit of an exaggeration, but a defensible one.

The turning-point in the evolution of group theory was the work of a young Frenchman, Évariste Galois. There was a long and complicated pre-history – Galois's ideas did not arise from a vacuum. And there was an equally complicated and often somewhat muddled post history, as mathematicians experimented with the new concept, and tried to work out what was important and what was not. But it was Galois, more than anyone else, who understood clearly the need for groups, worked out some of their fundamental features, and demonstrated their value in core mathematics. Not entirely surprisingly, his work went almost unnoticed during his lifetime. It was a little too original, perhaps, but it has to be said that Galois's personality, and his fierce involvement in revolutionary politics, did not help. He was a tragic figure living in a time of many personal tragedies, and his life was one of the more dramatic, and perhaps romantic, among those of major mathematicians.

Solving equations

The story of group theory goes right back to the ancient Babylonian work on quadratic equations. As far as the Babylonians were concerned, their method was intended for practical use; it was a computational technique, and they seem not to have asked deeper questions about it. If you knew how to find square roots, and had mastered basic arithmetic, you could solve quadratics.

The symmetries of a quadratic

Consider a quadratic equation, in the slightly simplified form

$$x^2+px+q = 0$$

Suppose that the two solutions are $x = a$ and $x = b$

$$x^2+px+q = (x-a)\ (x-b)$$

Then this tells us that

$$a + b = -p \qquad ab = q$$

So although we don't yet know the solutions, we do know their sum and their product – without doing any serious work.

Why is this? The sum $a + b$ is the same as $b + a$ – it does not change when the solutions are permuted. The same goes for $ab = ba$. It turns out that *every* symmetric function of the solutions can be expressed in terms of the coefficients p and q. Conversely, any expression in p and q is always a symmetric function of a and b. Taking a broad view, the connection between the solutions and the coefficients is determined by a symmetry property.

Asymmetric functions do not behave like this. A good example is the difference $a - b$. When we swap a and b, this becomes $b - a$, which is different. However – the crucial observation – it is not *very* different. It is what we get from $a - b$ by changing its sign. So the square $(a - b)^2$ is fully symmetric. But any fully symmetric function of the solutions must be some expression in the coefficients. Take the square root, and we have expressed $a - b$ in terms of the coefficients, using nothing more esoteric than a square root. We already know $a + b$ – it is equal to $-p$. Since we also know $a - b$, the sum of these two numbers is $2a$ and the difference is $2b$. Dividing by 2, we obtain formulas for a and for b.

What we've done is to prove that there must *exist* a formula for the solutions a and b involving nothing more esoteric than a square root, based on general features of the symmetries of algebraic expressions. This is impressive: we have proved that the problem possesses a solution, without going to the bother of working out all the messy details that tell us what it is. In a sense, we have tracked down *why* the Babylonians were able to find a method. This little story puts the word 'understand' in a new light. You can understand *how* the Babylonian method produces a solution, by working through the steps and checking the logic. But now we have understood why there had to be some such method – not by exhibiting a solution, but by examining *general* properties of the presumed solutions. Here, the key property turned out to be symmetry.

With a bit more work, leading to an explicit expression for $(a - b)^2$, this method yields a formula for the solutions. It is equivalent to the formula that we learn in school, and to the method used by the Babylonians.

There are a few hints, in surviving clay tablets, that the Babylonians also thought about cubic equations, even some quartic equations. The Greeks, and after them the Arabs, discovered geometric methods for solving cubic equations based on conic sections. (We now know that the traditional Euclidean lines and circles cannot solve such problems exactly. Something more sophisticated was necessary; as it happens, conics do the job.) One of the prominent figures here was the Persian, Omar Khayyam. Omar solved all possible types of cubic by systematic geometric methods. But, as we have seen, an algebraic solution of cubic and quartic equations had to wait until the Renaissance, with the work of Del Ferro, Tartaglia, Fior, Cardano and his pupil Ferrari.

The pattern that seemed to be emerging from all this work was straightforward, even though the details were messy. You can solve any cubic using arithmetical operations, plus square roots, plus cube roots. You can solve any quartic using arithmetical operations, plus square roots, plus cube roots, plus fourth roots – though the latter can be reduced to two square roots taken in succession. It seemed plausible that this pattern would continue, so that you could solve any quintic using arithmetical operations plus square roots, cube roots, fourth roots and fifth roots. And so on, for equations of any degree. No doubt the formulas would be very complicated, and finding them would be even more complicated, but few seem to have doubted that they existed.

As the centuries went by, with no sign of any such formulas being found, a few of the great mathematicians decided to take a closer look at the whole area, to work out what was really going on behind the scenes, unify the known methods and simplify them so that it became obvious why they worked. Then, they thought, it would just be a matter of applying the same general principles, and the quintic would give up its secret.

The most successful and most systematic work along these lines was carried out by Lagrange. He reinterpreted the classical formulas in terms of the solutions that were being sought. What mattered, he said, was how certain special algebraic expressions in those solutions behaved when the solutions themselves were permuted – rearranged. He knew that any fully symmetric expression – one that remained exactly the same no matter how the solutions were shuffled – could be expressed in terms of the coefficients of the equation, making it a known quantity. More interesting were expressions that only took on a few different values when the solutions were permuted. These seemed to hold the key to the whole issue of solving the equation.

Lagrange's well-developed sense of mathematical form and beauty told him that this was a major idea. If something similar could be developed for cubic and quartic equations, then he might discover how to solve the quintic.

Using the same basic idea, he found that partly symmetric functions of the solutions allowed him to reduce a cubic equation to a quadratic. The quadratic introduced a square root, and the reduction process could be sorted out using a cube root. Similarly, any quartic equation could be reduced to a cubic, which he called the resolvent cubic. So you could solve a quartic using square and cube roots to deal with the resolvent cubic and fourth roots to relate the answer to the solutions you wanted. In both cases, the answers were identical to the classical Renaissance formulas. They had to be, really – those were the answers. But now Lagrange knew *why* those were the answers, and better still, he knew why answers existed to be found.

He must have got quite excited at this stage of his research. Moving on to the quintic, and applying the same techniques, you expect to get a resolvent quartic – job done. But, presumably to his disappointment, he didn't get a resolvent quartic. He got a resolvent sextic – an equation of the sixth degree. Instead of making things simpler, his method made the quintic more complicated.

Was this a flaw in the method? Might something even cleverer solve the quintic? Lagrange seems to have thought so. He wrote that he hoped his new viewpoint would be useful to anyone trying to develop a way to solve quintics. It does not seem to have occurred to him that there might not *be* any such method; that his approach failed because in general, quintics do not *have* solutions in 'radicals' – which are expressions involving arithmetical operations and various roots, such as fifth roots. To confuse things, *some* quintics do have such solutions, for instance, $x^5 - 2 = 0$ has the solution $x = \sqrt[5]{2}$. But that is a rather simple case, and not really typical.

All quintic equations *have* solutions, by the way; in general these are complex numbers, and they can be found numerically to any accuracy. The problem was about algebraic formulas for the solutions.

Search for a solution

As Lagrange's ideas started to sink in, there was a growing feeling that perhaps the problem could not be solved. Perhaps the general quintic equation cannot be solved by radicals. Gauss seems to have thought so, privately, but expressed the view that this was not a problem he thought was worth tackling. It is perhaps one of the few instances where his intuition about what is important let him down; another was Fermat's Last Theorem, but here the necessary methods were beyond even Gauss, and took a couple of centuries to emerge. But, ironically, Gauss had already initiated some of the necessary algebra to prove the insolubility of the quintic. He had introduced it in his work on the construction of regular polygons by ruler and compass. And he had also, in this work, set a precedent, by proving (to his own satisfaction, at any rate) that some polygons could not be constructed in that manner. The regular 9-gon was an example. Gauss knew this, but never wrote down a proof; one was supplied a little later by Pierre Wantzel. So Gauss had established a precedent for the proposition that some problems might not be soluble by particular methods.

The first person to attempt a proof of the impossibility was Paolo Ruffini, who became a mathematics professor at the University of Modena in 1789. By pursuing Lagrange's ideas about symmetric functions, Ruffini became convinced that there is no formula, involving nothing more esoteric than nth roots, to solve the quintic. In his *General Theory of Equations* of 1799 he claimed a proof that 'The algebraic solution of general equations of degree greater than four is always impossible'. But the proof was so long – 500 pages – that no one was willing to check it, especially since there were rumours of mistakes. In 1803 Ruffini published a new, simplified proof, but it fared no better. During his lifetime, Ruffini never managed to secure the credit for proving that the quintic is insoluble.

Ruffini's most important contribution was the realization that permutations can be combined with each other. Until then, a permutation was a rearrangement of some collection of symbols. For instance, if we number the roots of a quintic as 12345, then these symbols can be rearranged as 54321, or 42153, or 23154 or whatever. There are 120 possible arrangements. Ruffini realized that such a rearrangement could be viewed in another way: as a recipe for rearranging *any other set* of five symbols. The trick was to compare the standard order 12345 with the rearranged order. As a simple example, suppose the rearranged order was 54321. Then the rule for getting from the initial standard order to the new order was simple: reverse the order. But you can reverse the order of *any* sequence of five symbols. If the symbols are *abcde*, the reverse is *edcba*. If the symbols start out as 23451, then their reverse is 15432. This new way of viewing a permutation meant that you could perform two permutations in turn – a kind of multiplication of permutations. The algebra of permutations, multiplied in this way, held the key to the secrets of the quintic.

Abel

We now know that there was a technical error in Ruffini's proof, but the main ideas are sound and the gap can be filled. He did achieve one thing: his book led to a vague but widespread feeling that the quintic is not soluble by radicals. Hardly anyone thought that Ruffini had *proved* this, but mathematicians began to doubt that a solution could exist. Unfortunately the main effect of this belief was to dissuade anyone from working on the problem.

An exception was Abel, a young Norwegian with a precocious talent for mathematics, who thought that he had solved the quintic while still at school. He eventually discovered a mistake, but remained intrigued by the problem, and kept working on it intermittently. In 1823 he found a proof of the impossibility of solving the quintic, and this proof was completely correct. Abel used a similar strategy to Ruffini's but his tactics were better. At first he was unaware of Ruffini's research; later he clearly knew of it, but he stated that it was incomplete. However, he did not refer to any specific problem with Ruffini's proof. Ironically, one step in Abel's proof is exactly what is needed to fill the gap in Ruffini's.

We can get a general idea of Abel's methods without going into too many technicalities. He set up the problem by distinguishing two kinds of algebraic operation. Suppose we start with various quantities – they may be specific numbers or algebraic expressions in various unknowns. From them we can build many other quantities. The easy way to do this is to combine the existing quantities by adding them, subtracting them, multiplying them or dividing them. So from a simple unknown, x, we can create expressions like x^2, $3x + 4$ or $\frac{x+7}{2x-3}$. Algebraically, all of these expressions are on much the same footing as x itself.

The second way to get new quantities from existing ones is to use radicals. Take one of the above mentioned harmless modifications of existing quantities, and extract some root. Call such a step adjoining a radical. If it is a square root, say that the degree of the radical is 2, if a cube root, then the degree is 3, and so on.

In these terms, Cardano's formula for the cubic can be summarized as the result of a two-step procedure. Start with the coefficients of the cubic (and any harmless combination of them). Adjoin a radical of degree 2. Then adjoin a further radical of degree 3. That's it. This description tells us what *kind* of formula arises, but not exactly what it is. Often the key to answering a mathematical riddle is not to focus on fine details, but to look at broad features. Less can mean more. When it works, this trick is spectacular, and here it worked beautifully. It allowed Abel to reduce any hypothetical formula for solving the quintic to its essential steps: extract some sequence of radicals, in some order, with various degrees. It is always possible to arrange for the degrees to be prime – for instance, a sixth root is the cube root of a square root.

Call such a sequence a *radical tower*. An equation is soluble by radicals if at least one of its solutions can be expressed by a radical tower. But instead of trying to find a radical tower, Abel merely assumed that there was a radical tower, and asked what the original equation must look like.

Without realizing it, Abel now filled the gap in Ruffini's proof. He showed that whenever an equation can be solved by radicals, there must exist a radical tower leading to that solution, involving only the coefficients of the original equation. This is called the Theorem on Natural Irrationalities and it states that nothing can be gained by including a whole pile of new quantities, unrelated to the original coefficients. This ought to be obvious, but Abel realized that it is in many ways the crucial step in the proof.

The key to Abel's impossibility proof is a clever preliminary result. Suppose we take some expression in the solutions x_1, x_2, x_3, x_4, x_5 of the equation, and extract its pth root for some prime number p. Moreover, assume that the original expression is unchanged when we apply two special permutations

$$S: x_1, x_2, x_3, x_4, x_5 \rightarrow x_2, x_3, x_1, x_4, x_5$$

and

$$T: x_1, x_2, x_3, x_4, x_5 \rightarrow x_1, x_2, x_4, x_5, x_3.$$

Then, Abel showed, the pth root of that expression is *also* unchanged when we apply S and T. This preliminary result leads directly to the proof of the impossibility theorem, by 'climbing the tower' step by step. Assume that the quintic can be solved by radicals, so there is a radical tower that starts with the coefficients and climbs all the way to some solution.

The first floor of the tower – the harmless expressions in the coefficients – is unchanged when we apply the permutations S and T, because those permute the solutions, not the coefficients. Therefore, by Abel's preliminary result, the second floor of the tower is *also* unchanged when we apply S and T, because it is reached by adjoining a pth root of something in the ground floor, for some prime p. By the same reasoning, the third floor of the tower is unchanged when we apply S and T. So is the fourth floor, the fifth floor, ..., all the way to the top floor.

However, the top floor contains some solution of the equation. Could it be x_1? If so, then x_1 must be unchanged when we apply S. But S applied to x_1 gives x_2, not x_1, so that's no good. For similar reasons, sometimes using T, the solution defined by the tower cannot be x_2, x_3, x_4 or x_5 either. *All* five solutions are excluded from any such tower – so the hypothetical tower cannot, in fact, contain a solution.

There is no escape from this logical trap. The quintic is unsolvable because any solution (by radicals) must have self-contradictory properties, and therefore cannot exist.

Galois

The pursuit not just of the quintic, but of all algebraic equations, was now taken up by Évariste Galois, one of the most tragic figures in the history of mathematics. Galois set himself the task of determining which equations could be solved by radicals, and which could not. Like several of his predecessors, he realized that the key to the algebraic solution of equations was how the solutions behaved when permuted. The problem was about symmetry.

Ruffini and Abel had realized that an expression in the solutions did not have to be either symmetric or not. It could be partially symmetric: unchanged by some permutations but not by others. Galois noticed that the permutations that fix some expression in the roots do not form any old collection. They have a simple, characteristic feature. If you take any two permutations that fix the expression, and multiply them together, the result *also* fixes the expression. He called such a system of permutations a *group*. Once you've realized that this is true, it is very easy to prove it. The trick is to notice it, and to recognize its significance.

Évariste Galois

1811-1832

Évariste Galois was the son of Nicholas Gabriel Galois and Adelaide Marie Demante. He grew up in revolutionary France, developing distinctly left-wing political views. His great contribution to mathematics went unrecognized until 14 years after his death.

The French revolution had begun with the storming of the Bastille in 1789 and the execution of Louis XVI in 1793. By 1804 Napoleon Bonaparte had proclaimed himself Emperor, but after a series of military defeats he was forced to abdicate, and the monarchy was restored in 1814 under Louis XVIII. By 1824, Louis had died and the king was now Charles X.

In 1827 Galois began to display an unusual talent for—and an obsession with—mathematics. He tried to gain entrance to the prestigious École Polytechnique, but failed the examination. In 1829 his father, then town Mayor, hanged himself when his political enemies invented a phoney scandal. Shortly after, Galois tried once more to enter the École Polytechnique, and failed again. Instead, he went to the École Normale.

In 1830 Galois submitted his researches on the solution of algebraic equations for a prize offered by the Academy of Sciences. The referee, Fourier, promptly died, and the paper was lost. The prize went to Abel (who was by then dead of tuberculosis) and to Carl Jacobi. In the same year Charles X was deposed and fled for his life. The director of the École Normale locked his students in to prevent them joining in. Galois, furious, wrote a sarcastic letter attacking the director for cowardice, and was promptly expelled.

As a compromise, Louis-Phillipe was made king. Galois joined a republican militia, the Artillery of the National Guard, but the new king abolished it. Nineteen of the Guard's officers were arrested and tried for sedition, but the jury threw the charges out, and the Guard held a dinner to celebrate. Galois proposed an ironic toast to the king, holding a knife in his hand. He was arrested, but acquitted because (so he claimed) the toast had been 'To Louis-Phillipe, if he betrays', and not a threat to the king's life. But on Bastille Day Galois was arrested again, for wearing the now illegal uniform of the Guard.

In prison, he heard what had happened to his paper. Poisson had rejected it for being insufficiently clear. Galois tried to kill himself, but the other prisoners stopped him. His hatred of officialdom now became extreme, and he displayed signs of paranoia. But when a cholera epidemic began, the prisoners were released.

At that point Galois fell in love with a woman whose name was for many years a mystery; she turned out to have been Stephanie du Motel, the daughter of the doctor in Galois's lodgings. The affair did not prosper, and Stephanie ended it. One of Galois's revolutionary comrades then challenged him to a duel, apparently over Stephanie. A plausible theory, advanced by Tony Rothman, is that the opponent was Ernest Duchâtelet, who had been imprisoned along with Galois. The duel seems to have been a form of Russian roulette, involving a random choice from two pistols, only one being loaded, and fired at point blank range. Galois chose the wrong pistol, was shot in the stomach, and died the next day.

The night before the duel he wrote a long summary of his mathematical ideas, including a description of his proof that all equations of degree 5 or higher cannot be solved by radicals. In this work he developed the concept of a group of permutations, and took the first important steps towards group theory. His manuscript was nearly lost, but it made its way to Joseph Liouville, a member of the Academy. In 1843 Liouville addressed the Academy, saying that in Galois's papers he had found a solution 'as correct as it is deep of this lovely problem: given an irreducible equation of prime degree, decide whether or not it is soluble by radicals'. Liouville published Galois's papers in 1846, finally making them accessible to the mathematical community.

The upshot of Galois's ideas is that the quintic cannot be solved by radicals because it has *the wrong kind of symmetries*. The group of a general quintic equation consists of all permutations of the five solutions. The algebraic structure of this group is inconsistent with a solution by radicals.

Galois worked in several other areas of mathematics, making equally profound discoveries. In particular he generalized modular arithmetic to classify what we now call *Galois fields*. These are finite systems in which the arithmetical operations of addition, subtraction, multiplication and division can be defined, and all the usual laws apply. The size of a Galois field is always a power of a prime, and there is exactly one such field for each prime power.

Jordan

The concept of a group first emerged in a clear form in the work of Galois, though with earlier hints in Ruffini's epic writings and the elegant researches of Lagrange. Within a decade of Galois's ideas becoming widely available, thanks to Liouville, mathematics was in possession of a well-developed *theory* of groups. The main architect of this theory was Camille Jordan, whose 667-page work *Traité de Substitutions et des Équations Algébriques* was published in 1870. Jordan developed the entire subject in a systematic and comprehensive way.

Jordan's involvement with group theory began in 1867, when he exhibited the deep link with geometry in a very explicit manner, by classifying the basic types of motion of a rigid body in Euclidean space. More importantly, he made a very good attempt to classify how these motions could be combined into groups. His main motivation was the crystallographic research of Auguste Bravais, who initiated the mathematical study of crystal symmetries, especially the underlying atomic lattice. Jordan's papers generalized the work of Bravais. He announced his classification in 1867, and published details in 1868–9.

What group theory did for them

One of the first serious applications of group theory to science was the classification of all possible crystal structures. The atoms in a crystal form a regular three-dimensional lattice, and the main mathematical point is to list all possible symmetry groups of such lattices, because these effectively form the symmetries of the crystal.

In 1891 Evgraf Fedorov and Arthur Schönflies proved that there are exactly 230 distinct crystallographic space groups. William Barlow obtained a similar but incomplete list.

Modern techniques for finding the structure of biological molecules, such as proteins, rely on passing X-rays through a crystal formed by that molecule and observing the resulting diffraction patterns. The symmetries of the crystal are important in deducing the shape of the molecule concerned. So is Fourier analysis.

Technically, Jordan dealt only with *closed* groups, in which the limit of any sequence of motions in the group is also a motion in the same group. These include all finite groups, for trivial reasons, and also groups like all rotations of a circle about its centre. A typical example of a non-closed group, not considered by Jordan, might be all rotations of a circle about its centre through rational multiples of 360°. This group exists, but does not satisfy the limit property (because, for example, it fails to include rotation by $360 \times \sqrt{2}$ degrees, since $\sqrt{2}$ is not rational). The non-closed groups of motions are enormously varied and almost certainly beyond any sensible classification. The closed ones are tractable, but difficult.

The main rigid motions in the plane are translations, rotations, reflections and glide reflections. In three-dimensional space, we also encounter *screw* motions, like the movement of a corkscrew: the object translates along a fixed axis and simultaneously rotates about the same axis.

Jordan began with groups of translations, and listed ten types, all mixtures of continuous translations (by any distance) in some directions and discrete translations (by integer multiples of a fixed distance) in other directions. He also listed the main finite groups of rotations and reflections: cyclic, dihedral, tetrahedral, octahedral and icosahedral. He distinguished the group O(2) of all rotations and reflections that leave a line in space, the *axis*, fixed, and the group O(3) of all rotations and reflections that leave a point in space, the *centre*, fixed.

Later it became clear that his list was incomplete. For instance, he had missed out some of the subtler crystallographic groups in three-dimensional space. But his work was a major step towards the understanding of Euclidean rigid motions, which are important in mechanics, as well as in the main body of pure mathematics.

Jordan's book is truly vast in scope. It begins with modular arithmetic and Galois fields, which as well as providing examples of groups also constitute essential background for everything else in the book. The middle third deals with groups of permutations, which Jordan calls substitutions. He sets up the basic ideas of normal subgroups which are what Galois used to show that the symmetry group of the quintic is inconsistent with a solution by radicals, and proves that these subgroups can be used to break a general group into simpler pieces. He proves that the sizes of these pieces do not depend on how the original group is broken up. In 1889 Otto Hölder improved this result, interpreting the pieces as groups in their own right, and proved that their group structure, not just their size, is independent of how the group is broken up. Today this result is called the Jordan–Hölder Theorem.

A group is *simple* if it does not break up in this way. The Jordan–Hölder Theorem effectively tells us that the simple groups relate to general groups in the same way that atoms relate to molecules in chemistry. Simple groups are the atomic constituents of all groups. Jordan proved that the alternating group A_n, comprising all permutations of n symbols that switch an even number of pairs of symbols, is simple whenever $n \geq 5$. This is the main group-theoretic reason why the quintic is insoluble by radicals.

A major new development was Jordan's theory of linear substitutions. Here the transformations that make up the group are not permutations of a finite set, but linear changes to a finite list of variables. For example, three variables x, y, z might transform into new variables X, Y, Z by way of linear equations

$$X = a_1x + a_2y + a_3z$$

$$Y = b_1x + b_2y + b_3z$$

$$Z = c_1x + c_2y + c_3z$$

where the a's, b's and c's are constants. To make the group finite, Jordan usually took these constants to be elements of the integers modulo some prime, or more generally a Galois field.

Also in 1869, Jordan developed his own version of Galois theory and included it in the *Traité*. He proved that an equation is soluble if and only if its group is soluble, which means that the simple components all have prime order. He applied Galois's theory to geometric problems.

Symmetry

The 4000-year-old quest to solve quintic algebraic equations was brought to an abrupt halt when Ruffini, Abel and Galois proved that no solution by radicals is possible. Although this was a negative result, it had a huge influence on the subsequent development of both mathematics and science. This happened because the method introduced to prove the impossibility turned out to be central to the mathematical understanding of symmetry, and symmetry turned out to be vital in both mathematics and science.

What group theory does for us

Group theory is now indispensable throughout mathematics, and its use in science is widespread. In particular, it turns up in theories of pattern formation in many different scientific contexts.

One example is the theory of reaction–diffusion equations, introduced by Alan Turing in 1952 as a possible explanation of symmetric patterns in the markings of animals. In these equations, a system of chemicals can diffuse across a region of space, and the chemicals can also react to produce new chemicals. Turing suggested that some such process might set up a pre-pattern in a developing animal embryo, which later on could be turned into pigments, revealing the pattern in the adult.

Suppose for simplicity that the region is a plane. Then the equations are symmetric under all rigid motions. The only solution of the equations that is symmetric under all rigid motions is a uniform state, the same everywhere. This would translate into an animal without any specific markings, the same colour all over. However, the uniform state may be unstable, in which case the actual solution observed will be symmetric under some rigid motions but not others. This process is called *symmetry-breaking*.

A typical symmetry-breaking pattern in the plane consists of parallel stripes. Another is a regular array of spots. More complicated patterns are also possible. Interestingly, spots and stripes are among the commonest patterns in animal markings, and many of the more complicated mathematical patterns are also found in animals. The actual biological process, involving genetic effects, must be more complicated than Turing assumed, but the underlying mechanism of symmetry-breaking must be mathematically very similar.

The effects were profound. Group theory led to a more abstract view of algebra, and with it a more abstract view of mathematics. Although many practical scientists initially opposed the move towards abstraction, it eventually became clear that abstract methods are often more powerful than concrete ones, and most opposition has disappeared. Group theory also made it clear that negative results may still be important, and that an insistence on proof can sometimes lead to major discoveries. Suppose that mathematicians had simply assumed without proof that quintics cannot be solved, on the plausible grounds that no one could find a solution. Then no one would have invented group theory to explain why they cannot be solved. If mathematicians had taken the easy route, and assumed the solution to be impossible, mathematics and science would have been a pale shadow of what they are today.

That is why mathematicians insist on proofs.

CHAPTER 14

Algebra Comes of Age

Numbers give way to structures

By 1860 the theory of permutation groups was well developed. The theory of invariants – algebraic expressions that do not change when certain changes of variable are performed – had drawn attention to various infinite sets of transformations, such as the *projective group* of all projections of space. In 1868 Camille Jordan had studied groups of motions in three-dimensional space, and the two strands began to merge.

Sophisticated concepts

A new kind of algebra began to appear, in which the objects of study were not unknown numbers, but more sophisticated concepts: permutations, transformations, matrices. Last year's processes had become this year's things. The long-standing rules of algebra often had to be modified to fit the needs of these new structures. Alongside groups, mathematicians started to study structures called rings and fields, and a variety of algebras.

One stimulus to this changing vision of algebras came from partial differential equations, mechanics and geometry: the development of Lie groups and Lie algebras. Another source of inspiration was number theory: here algebraic numbers could be used to solve Diophantine equations, understand reciprocity laws and even to attack Fermat's Last Theorem. Indeed, the culmination of such efforts was the proof of Fermat's Last Theorem by Andrew Wiles in 1995.

Lie and Klein

In 1869 the Norwegian mathematician Sophus Lie became friendly with the Prussian mathematician Felix Klein. They had a common interest in line geometry, an offshoot of projective geometry introduced by Julius Plücker. Lie conceived a highly original idea: that Galois's theory of algebraic equations should have an analogue for differential equations. An algebraic equation can be solved by radicals only if it has the right kind of symmetries – that is, it has a soluble Galois group. Analogously, Lie suggested, a differential equation can be solved by classical methods only when the equation remains unchanged by a family of continuous transformations. Lie and Klein worked on variations of this idea during1869–1870; it culminated in 1872 in Klein's characterization of geometry as the invariants of a group, laid down in his Erlangen programme.

This programme grew out of a new way of thinking about Euclidean geometry, in terms of its symmetries. Jordan had already pointed out that the symmetries of the Euclidean plane are rigid motions of several kinds: translations, which slide the plane in some direction; rotations, which turn it about some fixed point; reflections, which flip it over about a fixed line; and, less obviously, glide reflections, which reflect it and then translate it in a direction perpendicular to the mirror line. These transformations form a group, the *Euclidean group*, and they are rigid in the sense that they do not change distances. Therefore they also do not change angles. Now lengths and angles are the basic concepts of Euclid's geometry. So Klein realized that these concepts are the invariants of the Euclidean group, the quantities that do not change when a transformation from the group is applied.

In fact, if you know the Euclidean group, you can deduce its invariants, and from these you get Euclidean geometry.

The same goes for other kinds of geometry. Elliptic geometry is the study of the invariants of the group of rigid motions in a positively curved space, hyperbolic geometry is the study of the invariants of the group of rigid motions in a negatively curved space, and projective geometry is the study of the invariants of the group of projections and so on. Just as coordinates relate algebra to geometry, so invariants relate group theory to geometry. Each geometry defines a corresponding group, the group of all transformations that preserve the relevant geometric concepts. Conversely, every group of transformations defines a corresponding geometry, that of the invariants.

Klein used this correspondence to prove that certain geometries were essentially the same as others, because their groups were identical except for interpretation. The deeper message is that any geometry is defined by its symmetries. There is one exception: Riemann's geometry of surfaces, the curvature of which can change from point to point. It does not quite fit into Klein's programme.

Lie groups

Lie and Klein's joint research led Lie to introduce one of the most important ideas in modern mathematics, that of a continuous transformation group, now known as a *Lie group*. It is a concept that has revolutionized both mathematics and physics, because Lie groups capture many of the most significant symmetries of the physical universe, and symmetry is a powerful organizing principle – both for the underlying philosophy of how we represent nature mathematically, and for technical calculations.

Sophus Lie created the theory of Lie groups in a flurry of activity beginning in the autumn of 1873. The concept of a Lie group has evolved considerably since Lie's early work. In modern terms, a Lie group is a structure having both algebraic and topological

Felix Klein

1849–1925

Klein was born in Düsseldorf to an upper-class family – his father was secretary to the head of the Prussian government. He went to Bonn University, planning to become a physicist, but he became laboratory assistant to Julius Plücker. Plücker was supposed to be working in mathematics and experimental physics, but his interests had focused on geometry, and Klein fell under his influence. Klein's 1868 thesis was on line geometry as applied to mechanics.

By 1870 he was working with Lie on group theory and differential geometry. In 1871 he discovered that non-Euclidean geometry is the geometry of a projective surface with a distinguished conic section. This fact proved, very directly and obviously, that non-Euclidean geometry is logically consistent if Euclidean geometry is. This pretty much ended the controversy over the status of non-Euclidean geometry.

In 1872 Klein became professor at Erlangen, and in his Erlangen programme of 1872 he unified almost all known types of geometry, and clarified links between them, by considering geometry as the invariants of a transformation group. Geometry thereby became a branch of group theory. He wrote this article for his inaugural address, but did not actually present it on that occasion. Finding Erlangen uncongenial, he moved to Munich in 1875. He married Anne Hegel, granddaughter of the famous philosopher. Five years later he went to Leipzig, where he blossomed mathematically.

Klein believed that his best work was in the theory of complex functions, where he made deep studies of functions invariant under various groups of transformations of the complex plane. In particular he developed the theory of the simple group of order 168 in this context. He engaged in rivalry with Poincaré to solve the uniformization problem for complex functions, but his health collapsed, possibly because of the strenuous effort involved.

> In 1886 Klein was made professor at the University of Göttingen, and concentrated on administration, building one of the best schools of mathematics in the world. He remained there until retirement in 1913.

properties, the two being related. Specifically, it is a group (a set with an operation of composition that satisfies various algebraic identities, most notably the associative law) and a topological manifold (a space that locally resembles Euclidean space of some fixed dimension but which may be curved or otherwise distorted on the global level), such that the law of composition is continuous (small changes in the elements being composed produce small changes in the result). Lie's concept was more concrete: a group of continuous transformations in many variables. He was led to study such transformation groups while seeking a theory of the solubility or insolubility of differential equations, analogous to that of Évariste Galois for algebraic equations, but today they arise in an enormous variety of mathematical contexts, and Lie's original motivation is not the most important application.

Perhaps the simplest example of a Lie group is the set of all rotations of a circle. Each rotation is uniquely determined by an angle between 0° and 360°. The set is a group because the composition of two rotations is a rotation – through the sum of the corresponding angles. It is a manifold of dimension one, because angles correspond one-to-one with points on a circle, and small arcs of a circle are just slightly bent line segments, a line being Euclidean space of dimension one. Finally, the composition law is continuous because small changes in the angles being added produce small changes in their sum.

A more challenging example is the group of all rotations of three-dimensional space that preserve a chosen origin. Each rotation is determined by an axis – a line through the origin in an arbitrary direction – and an angle of rotation about that axis. It takes two variables to determine an axis (say the latitude and longitude of the point in which it meets a reference sphere centred on the origin) and a third to determine the angle of rotation; therefore this group has dimension three. Unlike the group of rotations of a circle, it is non-commutative – the result of combining two transformations depends upon the order in which they are performed.

In 1873, after a detour into PDEs, Lie returned to transformation groups, investigating properties of infinitesimal transformations. He showed that infinitesimal transformations derived from a continuous group are *not* closed under composition, but they *are* closed under a new operation known as the *bracket*, written $[x,y]$. In matrix notation this is the commutator $xy - yx$ of x and y. The resulting algebraic structure is now known as a *Lie algebra*. Until about 1930 the terms Lie group and Lie algebra were not used: instead these concepts were referred to as continuous group and infinitesimal group respectively.

There are strong interconnections between the structure of a Lie group and that of its Lie algebra, which Lie expounded in a three-volume work *Theorie der Transformationsgruppen* (*Theory of Transformation Groups*) written jointly with Friedrich Engel. They discussed in detail four classical families of groups, two of which are the rotation groups in n-dimensional space for odd or even n. The two cases are rather different, which is why they are distinguished. For example, in odd dimensions a rotation always possesses a fixed axis; in even dimensions it does not.

Killing

The next really substantial development was made by Wilhelm Killing. In 1888 Killing laid the foundations of a structure theory for Lie algebras, and in particular he classified all the *simple* Lie algebras, the basic building blocks out of which all other Lie algebras are composed. Killing started from the known structure of the most straightforward simple Lie algebras, the *special linear* Lie algebras sl(n), for $n \geq 2$. Start with all $n \times n$ matrices with complex entries, and let the Lie bracket of two matrices A and B be $AB - BA$. This Lie algebra is not simple, but the sub-algebra, sl(n), of all matrices whose diagonal terms sum to zero, is simple. It has dimension $n^2 - 1$.

Killing knew the structure of this algebra, and he showed that any simple Lie algebra had a similar kind of structure. It is remarkable that he could prove something so specific, starting only with the knowledge that the Lie algebra is simple. His method was to associate to each simple Lie algebra a geometric structure known as a *root system*. He used methods of linear algebra to study and classify root systems, and then derived the structure of the corresponding Lie algebra from that of the root system. So classifying the possible root system geometries is effectively the same as classifying the simple Lie algebras.

The upshot of Killing's work is remarkable. He proved that the simple Lie algebras fall into four infinite families, now called A_n, B_n, C_n and D_n. Additionally, there were five exceptions: G_2, F_4, E_6, E_7 and E_8. Killing actually thought there were six exceptions, but two of them turned out to be the same algebra in two different guises. The dimensions of the exceptional Lie algebras are 12, 56, 78, 133 and 248. They remain a little mysterious, although we now understand fairly clearly why they exist.

Simple Lie groups

Because of the close connections between a Lie group and its Lie algebra, the classification of simple Lie algebras also led to a classification of the simple Lie groups. In particular the four families A_n, B_n, C_n and D_n are the Lie algebras of the four classical families of transformation groups. These are, respectively, the group of all linear transformations in $(n + 1)$-dimensional space, the rotation group in $(2n + 1)$-dimensional space, the symplectic group in $2n$ dimensions, which is important in classical and quantum mechanics and optics, and the rotation group in 2n-dimensional space. A few finishing touches to this story were added later; notably the introduction by Harold Scott MacDonald Coxeter and Eugene (Evgenii) Dynkin of a graphical approach to the combinatorial analysis of root systems, now known as *Coxeter* or *Dynkin diagrams*.

Lie groups are important in modern mathematics for many reasons. For example, in mechanics, many systems have symmetries, and those symmetries make it possible to find solutions of the dynamical equations. The symmetries generally form a Lie group. In mathematical physics, the study of elementary particles relies heavily upon the apparatus of Lie groups, again because of certain symmetry principles. Killing's exceptional group E_8 plays an important role in superstring theory, an important current approach to the unification of quantum mechanics and general relativity. Simon Donaldson's epic discovery of 1983 that four-dimensional Euclidean space possesses non-standard differentiable structures rests, fundamentally, on an unusual feature of the Lie group of all rotations in four-dimensional space. The theory of Lie groups is vital to the whole of modern mathematics.

Abstract groups

In Klein's Erlangen programme it is essential that the groups concerned consist of transformations; that is, the elements of the group act on some space. Much of the early work on groups assumed this structure. But further research required one extra piece of abstraction: to retain the group property but throw away the space. A group consisted of mathematical entities that could be combined to yield similar entities, but those entities did not have to be transformations.

Numbers are an example. Two numbers (integer, rational, real, complex) can be added, and the result is a number of the same kind. Numbers form a group under the operation of addition. But numbers are not transformations. So even though the role of groups as transformations had unified geometry, the assumption of an underlying space had to be thrown away to unify group theory.

Among the first to come close to taking this step was Arthur Cayley, in three papers of 1849 and 1854. Here Cayley said that a group comprises a set of *operators* $1, a, b, c$ and so on. The compound ab of any two operators must be another operator; the special operator 1 satisfies $1a = a$ and $a1 = a$ for all operators a; finally, the associative law $(ab)c = a(bc)$ must hold. But his operators still operated on something (a set of variables). Additionally, he had omitted a crucial property: that every a must have an inverse a' such that $a'a = aa' = 1$. So Cayley came close, but missed the prize by a whisker.

In 1858 Richard Dedekind allowed the group elements to be arbitrary entities, not just transformations or operators, but he included the commutative law $ab = ba$ in his definition. His idea was fine for its intended purpose, number theory, but it excluded most of the interesting groups in Galois theory, let alone the wider mathematical world. The modern concept of an abstract group was introduced by Walther van Dyck in 1882–3. He included the existence of an inverse, but rejected the need for the commutative law. Fully-fledged axiomatic treatments of groups were provided soon after, by Edward Huntington and Eliakim Moore in 1902, and by Leonard Dickson in 1905.

With the abstract structure of groups now separated from any specific interpretation, the subject developed rapidly. Early research consisted mostly of 'butterfly collecting' – people studied individual examples of groups, or special types, looking for common patterns. The main concepts and techniques appeared relatively quickly, and the subject thrived.

Number theory

Another major source of new algebraic concepts was number theory. Gauss started the process when he introduced what we now call *Gaussian integers*. These are complex numbers $a + bi$, where a and

b are integers. Sums and products of such numbers also have the same form. Gauss discovered that the concept of a prime number generalizes to Gaussian integers. A Gaussian integer is *prime* if it cannot be expressed as a product of other Gaussian integers in a non-trivial way. Prime factorization for Gaussian integers is unique. Some ordinary primes, such as 3 and 7, remain prime when considered as Gaussian integers, but others do not: for example $5 = (2 + i)(2 - i)$. This fact is closely connected with Fermat's theorem about primes and sums of two squares, and Gaussian integers illuminate that theorem and its relatives.

If we divide one Gaussian integer by another, the result need not be a Gaussian integer, but it comes close: it is of the form $a + bi$, where a and b are rational. These are the *Gaussian numbers*. More generally, number theorists discovered that something similar holds if we take any polynomial $p(x)$ with integer coefficients, and then consider all linear combinations $a_1x_1 + \cdots + a_nx_n$ of its solutions $x_1, \ldots, x_n$. Taking $a_1, \ldots, a_n$ to be rational, we obtain a system of complex numbers that is *closed* under addition, subtraction, multiplication and division – meaning that when these operations are applied to such numbers, the result is a number of the same kind. This system constitutes an *algebraic number field*. If instead we require $a_1, \ldots, a_n$, to be integers, the system is closed under addition, subtraction and multiplication, but not division: it is an *algebraic number ring*.

The most ambitious application of these new number systems was Fermat's Last Theorem: the statement that the Fermat equation, $x^n + y^n = z^n$, has no whole number solutions when the power n is three or more. Nobody could reconstruct Fermat's alleged 'remarkable proof' and it seemed increasingly doubtful that he had ever possessed one. However, some progress was made. Fermat found proofs for cubes and fourth powers, Peter Lejeune-Dirichlet dealt with fifth powers in 1828 and Henri Lebesgue found a proof for seventh powers in 1840.

In 1847 Gabriel Lamé claimed a proof for all powers, but Ernst Eduard Kummer pointed out a mistake. Lamé had assumed without proof that uniqueness of prime factorization is valid for algebraic numbers, but this is *false* for some (indeed most) algebraic number fields. Kummer showed that uniqueness fails for the algebraic number field that arises in the study of Fermat's Last Theorem for 23rd powers. But Kummer did not give up easily, and he found a way round this obstacle by inventing some new mathematical gadgetry, the theory of ideal numbers. By 1847 he had disposed of Fermat's Last Theorem for all powers up to 100, except for 37, 59 and 67. By developing extra machinery, Kummer and Dimitri Mirimanoff disposed of those cases too in 1857. By the 1980s similar methods had proved all cases up to the 150,000th power, but the method was running out of steam.

Rings, fields and algebras

Kummer's notion of an ideal number was cumbersome, and Dedekind reformulated it in terms of ideals, special subsystems of algebraic integers. In the hands of Hilbert's school at Göttingen, and in particular Emmy Noether, the entire area was placed on an axiomatic footing. Alongside groups, three other types of algebraic system were defined by suitable lists of axioms: rings, fields and algebras.

In a *ring*, operations of addition, subtraction and multiplication are defined, and they satisfy all the usual laws of algebra except for the commutative law of multiplication. If this law also holds, we have a *commutative ring*.

In a *field*, operations of addition, subtraction, multiplication and division are defined, and they satisfy all the usual laws of algebra including the commutative law of multiplication. If this law fails, we have a *division ring*.

An *algebra* is like a ring, but its elements can also be multiplied by various constants, the real numbers, complex numbers or – in the most general setting – by a field. The laws of addition are the usual

Emmy Amalie Noether

1882–1935

Emmy Noether was the daughter of the mathematician Max Noether and Ida Kaufmann, both of Jewish origin. In 1900 she qualified to teach languages but instead decided her future lay in mathematics. At that time German universities allowed women to study courses unofficially if the professor gave permission, and she did this from 1900 to 1902. Then she went to Göttingen, attending lectures by Hilbert, Klein and Minkowski in 1903 and 1904.

She gained a doctorate in 1907, under the invariant theorist Paul Gordan. Her thesis calculated a very complicated system of invariants. For men, the next step would be Habilitation, but this was not permitted for women. She stayed home in Erlangen, helping her disabled father, but she continued her research and her reputation quickly grew.

In 1915 she was invited back to Göttingen by Klein and Hilbert, who struggled to get the rules changed to allow her on to the faculty. They finally succeeded in 1919. Soon after her arrival she proved a fundamental theorem, often called Noether's Theorem, relating the symmetries of a physical system to conservation laws. Some of her work was used by Einstein to formulate parts of general relativity. In 1921 she wrote a paper on ring theory and ideals, taking an abstract axiomatic view. Her work formed a significant part of Bartel Leendert van der Waerden's classic text *Moderne Algebra*.

When Germany fell under Nazi rule she was dismissed because she was Jewish, and she left Germany to take up a position in the USA. Van der Waerden said that for her, 'relationships among numbers, functions and operations became transparent, amenable to generalization and productive only after they have been ... reduced to general conceptual relationships.'

ones, but the multiplication may satisfy a variety of different axioms. If it is associative, we have an associative algebra. If it satisfies some laws related to the commutator $xy - yx$, it is a Lie algebra.

There are dozens, maybe hundreds of different types of algebraic structure, each with its own list of axioms. Some have been invented just to explore the consequences of interesting axioms, but most arose because they were needed in some specific problem.

Finite simple groups

The high point of 20th century research on finite groups was the successful classification of all finite simple groups. This achieved for finite groups what Killing had achieved for Lie groups and their Lie algebras. Namely, it led to a complete description of all possible basic building blocks for finite groups, the *simple* groups. If groups are molecules, the simple groups are their constituent atoms.

Killing's classification of simple Lie groups proved that these must belong to one of four infinite families A_n, B_n, C_n and D_n, with exactly five exceptions, G_2, F_4, E_6, E_7 and E_8. The eventual classification of all finite simple groups was achieved by too many mathematicians to mention individually, but the overall programme for solving this problem was due to Daniel Gorenstein. The answer, published in 1888–90, is strangely similar: a list of infinite families, and a list of exceptions. But now there are many more families, and the exceptions number 26.

The families comprise the alternating groups (known to Galois) and a host of groups of Lie type which are like the simple Lie groups but over various finite fields rather than the complex numbers. There are some curious variations on this theme, too. The exceptions are 26 individuals, with hints of some common patterns, but no unified structure. The first proof that the classification is complete came from the combined work of hundreds of mathematicians, and its total length was around 10,000 pages. Moreover, some crucial parts of the proof were not published. Recent work by those remaining in this area of research has involved reworking the classification in a more streamlined manner, an approach made possible by already knowing the answer. The results are appearing as a series of textbooks, totalling around 2000 pages.

The most mysterious of the exceptional simple groups, and the largest, is the *monster*. Its order is

$$2^{46}\times3^{20}\times5^{9}\times7^{6}\times11^{2}\times13^{3}\times17\times19\times23\times29\times31\times41\times47\times59\times71$$

which equals

808017424794512875886459904961710757005754368000000000

and is roughly 8×10^{53}. Its existence was conjectured in 1973 by Bernd Fischer and Robert Griess. In 1980 Griess proved that it existed, and gave an algebraic construction as the symmetry group of a 196,884-dimensional algebra. The monster seems to have some unexpected links with number theory and complex analysis, stated by John Conway as the Monstrous Moonshine conjecture. This conjecture was proved by Richard Borcherds in 1992, and he was awarded a Fields Medal – the most prestigious award in mathematics – for it.

Fermat's Last Theorem

The application of algebraic number fields to number theory developed apace in the second half of the 20th century, and made contact with many other areas of mathematics, including Galois theory and algebraic topology. The culmination of this work was a proof of Fermat's Last Theorem, some 350 years after it was first stated.

The really decisive idea came from a beautiful area that lies at the heart of modern work on Diophantine equations: the theory of elliptic curves. These are equations in which a perfect square is equal to a cubic polynomial, and they represent the one area of Diophantine equations that mathematicians understand pretty well. However, the subject has its own big unsolved problems. The biggest of all is the Taniyama–Weil conjecture, named after Yutaka Taniyama and André Weil. This says that every elliptic curve can be represented in terms of modular functions – generalizations of trigonometric functions studied in particular by Klein.

What abstract algebra did for them

In his 1854 book *The Laws of Thought*, George Boole showed that algebra can be applied to logic, inventing what is now known as Boolean algebra.

Here we can do no more than convey a flavour of Boole's ideas. The most important logical operators are *not*, *and* and *or*. If a statement S is true, then not-S is false, and vice versa. S and T is true if, and only if, both S and T are true. S or T is true provided at least one of S, T is true – possibly both. Boole noticed that if we rewrite T as 1 and S as 0, then the algebra of these logical operations is very similar to ordinary algebra, provided we think of 0 and 1 as integers modulo 2, so that $1 + 1 = 0$ and $-S$ is the same as S. So not-S is $1+S$, S and T is ST, and S or T is $S+T+ST$. The sum $S+T$ corresponds to *exclusive* or (written *xor* by computer scientists). S *xor* T is true provided T is true or S is true, but not both. Boole discovered that his curious algebra of logic is entirely self-consistent if you bear its slightly weird rules in mind and use them systematically. This was one of the first steps towards a formal theory of mathematical logic.

Early in the 1980s Gerhard Frey found a link between Fermat's Last Theorem and elliptic curves. Suppose that a solution to Fermat's equation exists; then you can construct an elliptic curve with very unusual properties – so unusual that the curve's existence seems highly improbable. In 1986 Kenneth Ribet made this idea precise by proving that if the Taniyama–Weil conjecture is true, then Frey's curve cannot exist. Therefore the presumed solution of Fermat's equation cannot exist either, which would prove Fermat's Last Theorem. The approach depended on the Taniyama–Weil conjecture, but it showed that Fermat's Last Theorem is not just an isolated historical curiosity. Instead, it lies at the heart of modern number theory.

Andrew Wiles

1953 –

Andrew Wiles was born in 1953 in Cambridge, England. At the age of 10 he read about Fermat's Last Theorem and resolved to become a mathematician and prove it. By the time of his PhD he had pretty much abandoned this idea, because the theorem seemed so intractable, so he worked on the number theory of 'elliptic curves', an apparently different area. He moved to the USA and became a Professor at Princeton.

By the 1980s it was becoming clear that there might be an unexpected link between Fermat's Last Theorem and a deep and difficult question about elliptic curves. Gerhard Frey made this link explicit, by means of the so-called Taniyama–Shimura conjecture. When Wiles heard of Frey's idea he stopped all of his other work to concentrate on Fermat's Last Theorem, and after seven years of solitary research he convinced himself that he had found a proof, based on a special case of the Taniyama–Shimura Conjecture. This proof turned out to have a gap, but Wiles and Richard Taylor repaired the gap and a complete proof was published in 1995.

Other mathematicians soon extended the ideas to prove the full Taniyama–Shimura Conjecture, pushing the new methods further. Wiles received many honours for his proof, including the Wolf Prize. In 1998, being just too old for a Fields Medal, traditionally limited to people under 40, he was awarded a special silver plaque by the International Mathematical Union. He was made a Knight Commander of the Order of the British Empire in 2000.

Andrew Wiles, as a child, had dreamed of proving Fermat's Last Theorem, but when he became a professional he decided that it was just an isolated problem – unsolved, but not really important.

Ribet's work changed his mind. In 1993 he announced a proof of the Taniyama–Weil conjecture for a special class of elliptic curves, general enough to prove Fermat's Last Theorem. But when the paper was submitted for publication, a serious gap emerged. Wiles had almost given up when 'suddenly, totally unexpectedly, I had this incredible revelation ... it was so indescribably beautiful, it was so simple and so elegant, and I just stared in disbelief.' With the aid of Richard Taylor, he revised the proof and repaired the gap. His paper was published in 1995.

We can be sure that whatever ideas Fermat had in mind when he claimed to possess a proof of his Last Theorem, they must have been very different from the methods used by Wiles. Did Fermat really have a simple, clever proof, or was he deluding himself? It is a puzzle that, unlike his Last Theorem, may never be resolved.

Abstract mathematics

The move towards a more abstract view of mathematics was a natural consequence of the growing variety of its subject matter. When mathematics was mostly about numbers, the symbols of algebra were simply placeholders for numbers. But as mathematics grew, the symbols themselves started to take on a life of their own. The meaning of the symbols became less significant than the rules by which those symbols could be manipulated. Even the rules were not sacred: the traditional laws of arithmetic, such as the commutative law, were not always appropriate in new contexts.

It was not only algebra that became abstract. Analysis and geometry also focused on more general issues, for similar reasons. The main change in viewpoint occurred from the middle of the 19th century to the middle of the 20th. After that, a period of consolidation set in, as mathematicians tried to balance the conflicting needs of abstract formalism and applications to science. Abstraction and generality go hand in hand, but abstraction can also

What abstract algebra does for us

Galois fields form the basis of a coding system that is widely used in a variety of commercial applications, especially CDs and DVDs. Every time you play music, or watch a video, you are using abstract algebra.

These methods are known as *Reed–Solomon* codes, after Irving Reed and Gustave Solomon who introduced them in 1960. They are error-correcting codes based on a polynomial, with coefficients in a finite field, constructed from the data being encoded, such as the music or video signals. It is known that a polynomial of degree n is uniquely determined by its values at n distinct points. The idea is to calculate the polynomial at more than n points. If there are no errors, any subset of n data points will reconstruct the same polynomial. If not, then provided the number of errors is not too large, it is still possible to deduce the polynomial.

In practice the data are represented as encoded blocks, with $2^m - 1$ m-bit symbols per block, where a bit is a binary digit, 0 or 1. A popular choice is $m = 8$, because many of the older computers work in bytes – sequences of eight bits. Here the number of symbols in a block is 255. One common Reed–Solomon code puts 223 bytes of encoded data in each 255 byte block, using the remaining 32 bytes for parity symbols which state whether certain combinations of digits in the data should be odd or even. This code can correct up to 16 errors per block.

obscure the meaning of mathematics. But the issue is no longer whether abstraction is useful or necessary: abstract methods have proved their worth by making it possible to solve numerous long-standing problems, such as Fermat's Last Theorem. And what seemed little more than formal game-playing yesterday may turn out to be a vital scientific or commercial tool tomorrow.

CHAPTER 15

Rubber Sheet Geometry

Qualitative beats quantitative

The main ingredients of Euclid's geometry – lines, angles, circles, squares and so on – are all related to *measurement*. Line segments have lengths, angles are a definite size with 90° differing in important ways from 91° or 89°, circles are defined in terms of their radii, squares have sides of a given length. The hidden ingredient that makes all of Euclid's geometry work is length, a metric quantity, one which is unchanged by rigid motions and defines Euclid's equivalent concept to motion, congruence.

Topology

When mathematicians first stumbled across other types of geometry, these too were metric. In non-Euclidean geometry, lengths and angles are defined; they just have different properties from lengths and angles in the Euclidean plane. The arrival of projective geometry changed this: projective transformations can change lengths, and they can change angles. Euclidean geometry and the two main kinds of non-Euclidean geometry are rigid. Projective geometry is more flexible, but even here subtler invariants exist, and in Klein's picture what defines a geometry is a group of transformations and the corresponding invariants.

As the 19th century approached its end, mathematicians began to develop an even more flexible kind of geometry; so flexible, in fact, that it is often characterized as rubber-sheet geometry. More properly known as *topology*, this is the geometry of shapes that can be deformed or distorted in extremely convoluted ways. Lines can bend, shrink or stretch; circles can be squashed so that they turn into triangles or squares. All that matters here is *continuity*. The transformations allowed in topology are required to be continuous in the sense of analysis; roughly speaking, this means that if two points start sufficiently close together, they end up close together – hence the 'rubber sheet' image.

There is still a hint of metric thinking here: 'close together' is a metric concept. But by the early 20th century, even this hint had been removed, and topological transformations took on a life of their own. Topology quickly increased in status, until it occupied centre stage in mathematics – even though to begin with it seemed very strange, and virtually content-free. With transformations that flexible, what could be invariant? The answer, it turned out, was quite a lot. But the type of invariant that began to be uncovered was like nothing ever before considered in geometry. Connectivity – how many pieces does this thing have? Holes – is it all one lump, or are there tunnels through it? Knots – how is it tangled up, and can you undo the tangles? To a topologist, a doughnut and a coffee-cup are identical (but a doughnut and a tumbler are not); a doughnut is different from a round ball, but ball and tumbler are identical. An overhand knot is different from a figure-eight knot, but proving this fact required a whole new kind of machinery, and for a long time no one could prove that any knots existed at all.

It seems remarkable that anything so diffuse and plain weird could have any importance at all. But appearances are deceptive.

Continuity is one of the basic aspects of the natural world, and any deep study of continuity leads to topology. Even today we mostly use topology indirectly, as one technique among many. You don't find anything topological sitting in your kitchen – not obviously, at least. (However, you can occasionally find such items as a chaotic dishwasher, which uses the strange dynamics of two rotating arms to clean dishes more efficiently. And our understanding of the phenomenon of chaos rests on topology.) The main practical consumers of topology are quantum field theorists – a new use of the word 'practical' perhaps, but an important area of physics. Another application of topological ideas occurs in molecular biology, where describing and analysing the twists and turns of the DNA molecule requires topological concepts.

Behind the scenes, topology informs the whole of mainstream mathematics, and enables the development of other techniques with more obvious practical uses. It is a rigorous study of *qualitative* geometric features, as opposed to quantitative ones like lengths. This is why mathematicians consider topology to be of vast importance, whereas the rest of the world has hardly heard of it.

Polyhedra and the Königsberg Bridges

Although topology really only began to take off around 1900, it made an occasional appearance in earlier mathematics. Two items in the prehistory of topology were introduced by Euler: his formula for polyhedra and his solution of the puzzle of the Königsberg Bridges.

In 1639 Descartes had noticed a curious feature of the numerology of regular solids. Consider, for instance, a cube. This has six faces, 12 edges and eight vertices. Add 6 to 8 and you get 14, which is 2 larger than 12. How about a dodecahedron? Now there are 12 faces, 30 edges and 20 vertices. And $12 + 20 = 32$, which exceeds 30 by 2. The same goes for the tetrahedron, octahedron and icosahedron. In fact, the same relation seemed to work for almost

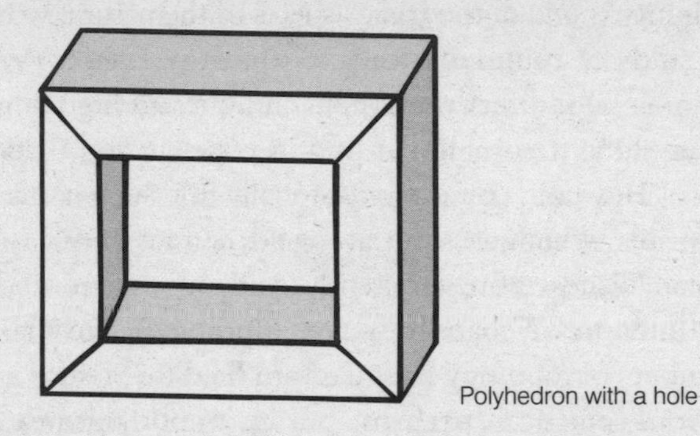

Polyhedron with a hole

any polyhedron. If a solid has F faces, E edges and V vertices, then $F + V = E + 2$, which we can rewrite as

$$F + V - E = 2$$

Descartes did not publish his discovery, but he wrote it down and his manuscript was read by Leibniz in 1675.

Euler was the first to publish this relationship, in 1750. He followed this up with a proof in 1751. He was interested in the relationship because he had been trying to classify polyhedra. Any general phenomenon such as this one had to be taken into account when performing such a classification.

Is the formula valid for *all* polyhedra? Not quite. A polyhedron in the form of a picture frame, with square cross-section and mitred corners, has 16 faces, 32 edges and 16 vertices, so that here $F + V - E = 0$. The reason for the discrepancy turns out to be the presence of a hole. In fact, if a polyhedron has g holes, then

$$F + V - E = 2 - 2g$$

What, exactly, is a hole? This question is harder than it looks. First, we are talking about the surface of the polyhedron, not its solid interior. In real life we make a hole in something by burrowing through its solid interior, but the above formulas make no reference to the interior of the polyhedron – just to the faces that constitute its surfaces, along with their edges and vertices. Everything we count lies on the surface here. Second, the only holes that change the numerology are those that burrow all the way through the polyhedron – tunnels with two ends, so to speak, not holes like workmen dig in the road. Third, such holes are not in the surface at all, though they are somehow delineated by it. When you buy a doughnut, you also buy its hole, but you can't buy a hole in its own right. It exists only by virtue of the doughnut's external surface, even though in this case you also buy the doughnut's solid interior.

It is easier to define what 'no holes' means. A polyhedron has no holes if it can be continuously deformed, creating curved faces and edges, so that it becomes (the surface of) a sphere. For these surfaces, $F + V - E$ really is always 2. And the converse holds as well: if $F + V - E = 2$ then the polyhedron can be deformed into a sphere.

The picture-frame polyhedron does not look as though it can be deformed into a sphere – where would the hole go? For a rigorous proof of the impossibility, we need look no further than the fact that for this polyhedron, $F + V - E = 0$. This relationship is impossible for surfaces deformable to spheres. So the numerology of polyhedra tells us significant features of their geometry, and those features can be topological invariants – unchanged by deformations.

Euler's formula is now viewed as a significant hint of a useful link between combinatorial aspects of polyhedra, such as numbers of faces, and topological aspects. In fact, it turns out to be easier to work backwards. To find out how many holes a surface has, work out $F + V - E - 2$, divide by 2, and change sign:

$$g = -(F + V - E - 2)/2$$

A curious consequence: we can now define how many holes a polyhedron has, without defining 'hole'.

An advantage of this procedure is that it is intrinsic to the

Cauchy's Proof of the Descartes–Euler Formula

Remove one face and stretch the surface of the solid out on a plane. This reduces F by 1, so we now prove that the resulting plane configuration of faces, lines and points has $F + V - E = 1$. To achieve this, first convert all faces to triangles by drawing diagonals. Each new diagonal leaves V unchanged, but increases both E and F by 1, so $F + V - E$ remains the same as before. Now start deleting edges, starting from the outside. Each such deletion reduces both F and E and leaves the number of vertices unchanged, so $F + V - E$ is once more unchanged. When you run out of faces to delete, you are left with a tree of edges and vertices, containing no closed loops. One by one, delete terminal vertices, along with the edge that meets them. Now E and V both decrease by 1, and again $F + V - E$ is unchanged. Eventually this process stops with one solitary vertex. Now $F = 0$, $E = 0$ and $V = 1$, so $F + V - E = 1$, as required.

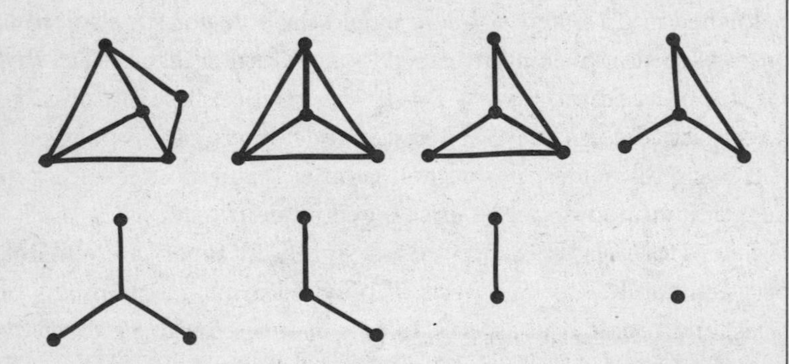

Example of Cauchy's proof

polyhedron. It does not involve visualizing the polyhedron in a surrounding three-dimensional space, which is how our eyes naturally see the hole. A sufficiently intelligent ant that lived on the surface of the polyhedron could work out that it has one hole, even if all it could see was the surface. This intrinsic point of view is natural in topology. It studies the shapes of things in their own right, not as part of something else.

At first sight, the problem of the Königsberg Bridges bears no relationship to the combinatorics of polyhedra. The city of Königsberg, then in Prussia, was situated on both banks of the river Pregelarme, in which there were two islands. The islands were linked to the banks, and to each other, by seven bridges. Apparently, the citizens of Königsberg had long wondered whether it was possible to take a Sunday stroll that would cross each bridge exactly once.

In 1735 Euler solved the puzzle; rather, he proved that there is no solution, and explained why. He made two significant contributions: he simplified the problem and reduced it to its bare essentials, and then he generalized it, to deal with all puzzles of a similar kind. He pointed out that what matters is not the size and shape of the islands, but how the islands, the banks and the bridges are connected. The whole problem can be reduced to a simple diagram of dots (vertices) joined by lines (edges), here shown superimposed on a map.

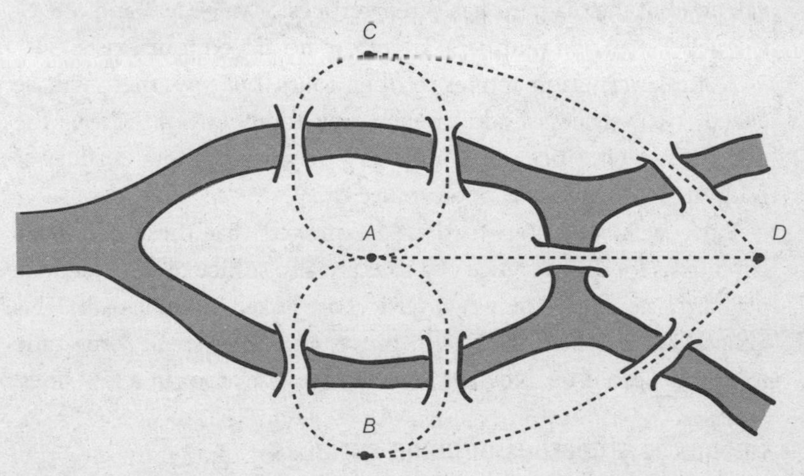

Problem of the Königsberg Bridges

To form this diagram, place one vertex on each land-mass – north bank, south bank, and the two islands. Join two vertices by an edge whenever a bridge exists that links the corresponding land masses. Here we end up with four vertices A, B, C, D and seven edges, one for each bridge.

The puzzle is then equivalent to a simpler one about the diagram. Is it possible to find a path – a connected sequence of edges – that includes each edge exactly once?

Euler used a symbolic description, but we can interpret his main ideas in terms of the diagram. He distinguished two types of path: an *open tour*, which starts and ends at different vertices, and a *closed tour*, which starts and ends at the same vertex. He proved that for this particular diagram neither kind of tour exists.

The key to the puzzle is to consider the *valency* of each vertex, that is, how many edges meet at that vertex. First, think of a closed tour. Here, every edge where the tour enters a vertex is matched by another, the next edge, along which the tour leaves that vertex. If a closed tour is possible, then the number of edges at any given vertex must therefore be even. In short, every vertex must have even valency. But the diagram has three vertices of valency 3 and one of valency 5 – all odd numbers. Therefore no closed tour exists.

A similar criterion applies to open tours, but now there will be exactly two vertices of odd valency: one at the start of the tour, the other at its end. Since the Königsberg diagram has four vertices of odd valency, there is no open tour either.

Euler went one step further: he proved that these necessary conditions for the existence of a tour are also sufficient, provided the diagram is connected (any two vertices are linked by some path). This general fact is a little trickier to prove, and Euler spent some time setting his proof up. Nowadays we can give a proof in a few lines.

Geometric properties of plane surfaces

Euler's two discoveries seem to belong to entirely different areas of mathematics, but on closer inspection they have elements in common. They are both about the combinatorics of polyhedral diagrams. One counts faces, edges and vertices and the other counts valencies; one is about a universal relation between three numbers, the other a relation that must occur if there exists a tour. But they are visibly similar in spirit. More deeply – and this went unappreciated for more than a century – both are invariant under continuous transformations. The positions of the vertices and edges do not matter: what counts is how they connect to each other. Both problems would look the same if the diagrams were drawn on a rubber sheet and the sheet were distorted. The only way to create significant differences would be to cut or tear the sheet, or glue bits together – but these operations destroy continuity.

The glimmerings of a general theory were apparent to Gauss, who from time to time made quite a lot of fuss about the need for some theory of the basic geometric properties of diagrams. He also developed a new topological invariant, which we now call the *linking number*, in work on magnetism. This number determines how one closed curve winds round another. Gauss gave a formula to calculate the linking number from analytic expressions for the curves. A similar invariant, the *winding number* of a closed curve relative to a point, was implicit in one of his proofs of the Fundamental Theorem of Algebra.

Gauss's main influence on the development of topology came from one of his students, Johann Listing, and his assistant Augustus Möbius. Listing studied under Gauss in 1834, and his work *Vorstudien zur Topologie* introduced the word topology. Listing himself would have preferred to call the subject geometry of position, but this phrase had been pre-empted by Karl von Staudt to mean projective geometry, so Listing found another word. Among other things, Listing looked for generalizations of Euler's formula for polyhedra.

It was Möbius who made the role of continuous transformations explicit. Möbius was not the most productive of mathematicians, but he tended to think everything through very carefully and very thoroughly. In particular, he noticed that surfaces do not always have two distinct sides, giving the celebrated *Möbius band* as an example. This surface was discovered independently by Möbius and Listing in 1858. Listing published it in *Der Census Räumlicher Complexe*, and Möbius put it in a paper on surfaces.

For a long time Euler's ideas on polyhedra were something of a side-issue in mathematics, but several prominent mathematicians began to glimpse a new approach to geometry, which they called 'analysis situs' – the analysis of position. What they had in mind was a *qualitative* theory of shape, in its own right, to supplement the more traditional quantitative theory of lengths, angles, areas and volumes. This view began to gain ground when issues of that kind emerged from traditional investigations in mainstream mathematics. A key step was the discovery of connections between complex analysis and the geometry of surfaces, and the innovator was Riemann.

The Riemann sphere

The obvious way to think of a complex function f is to interpret it as a mapping from one complex plane to another. The basic formula $w = f(z)$ for such a function tells us to take any complex number z, apply f to it and deduce another complex number w associated with z. Geometrically, z belongs to the complex plane, and w belongs to what is in effect a second, independent copy of the complex plane.

However, this point of view turns out not to be the most useful one, and the reason is *singularities*. Complex functions often have interesting points at which their normal, comfortable behaviour goes horribly wrong. For example, the function $f(z) = 1/z$ is well behaved for all z except zero. When $z = 0$, the value of the function is $1/0$, which makes no sense as an ordinary complex number, but can with some stretch of the imagination be thought of as infinity (symbol ∞). Specifically, if z gets very close to 0, then $1/z$ gets very big. Infinity in this sense is not a number, but

The Möbius Band

Topology has some surprises. The best known is the Möbius band (or Möbius strip), which can be formed by taking a long strip of paper, and joining the ends with a half-twist. Without this twist, we get a cylinder. The difference between these two surfaces becomes apparent if we try to paint them. We can paint the outer surface of a cylinder red and the inner surface blue. But if you start painting a Möbius band red on one side, and keep going until every part of the surface that connects to the red region has been covered, you end up coating the entire band in red paint. The inside surface connects to the outside thanks to that half-twist.

Another difference appears if you cut along the centre line of the band. It remains in one piece.

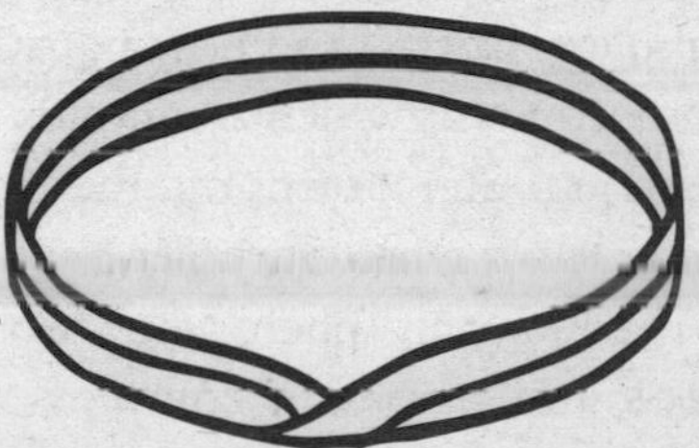

a term that describes a numerical process: become as large as you wish. Gauss had already noticed that infinities of this type create new types of behaviour in complex integration. They mattered.

Riemann found it useful to include ∞ among the complex numbers, and he found a beautiful geometric way to do this. Place a unit sphere so that it sits on top of the complex plane. Now associate points in the plane with points on the sphere by stereographic projection. That is, join the point in the plane to the North Pole of the sphere, and see where that line meets the sphere.

This construction is called the *Riemann sphere*. The new point at infinity is the North Pole of the sphere – the only point that does not correspond to a point in the complex plane. Amazingly, this

The Riemann sphere and the complex plane

construction fits beautifully into the standard calculations in complex analysis, and now equations like $1/0 = \infty$ make perfect sense. Points at which a complex function f takes the value ∞ are called *poles*, and it turns out that you can learn a great deal about f if you know where its poles lie.

The Riemann sphere alone would not have drawn attention to topological issues in complex analysis, but a second kind of singularity, called a *branch point*, made topology essential. The simplest example is the complex square root function, $f(z) = \sqrt{z}$. Most complex numbers have two distinct square roots, just like real numbers. These square roots differ only in their sign: one is minus the other. For example, the square roots of $2i$ turn out to be $1 + i$ and $-1 - i$, much as the real square roots of 4 are 2 and -2. However, there is one complex number with only *one* square root, namely 0. Why? Because $+0$ and -0 are equal.

To see why 0 is a branch point of the square root function imagine starting at the point 1 in the complex plane, and choosing one of the two square roots. The obvious choice is also 1. Now gradually move the point round the unit circle, and as you go, choose, for each position of the point, whichever of the two square roots keeps everything varying continuously. By the time you get half way round the circle, to -1, the square root has only gone a quarter of the way round, to $+i$, since $\sqrt{-1} = +i$ or $-i$. Continuing all the way round, we get back to the starting point 1. But the square root, moving at half the speed, ends up at -1. To return the square root to its initial value, the point has to go round the circle twice.

Riemann found a way to tame this kind of singularity, by doubling up the Riemann sphere into two layers. These layers are separate except at the points 0 and ∞, which is a second branch point. At these two points the layers merge – or, thinking about this the other way round, they branch from the single layers at 0 and ∞. Near these special points, the geometry of the layers is like a spiral staircase – with the unusual feature that if you climb two full turns up the staircase, you get back to where you started. The geometry of this surface tells us a lot about the square root function, and the same idea can be extended to other complex functions.

The description of the surface is rather indirect, and we can ask what shape is it? It is here that topology comes into play. We can continuously deform the spiral staircase description into something easier to visualize. The complex analysts found that, topologically, every Riemann surface is either a sphere, or a torus, or a torus with two holes, or a torus with three holes, etc. The number of holes, g, is known as the *genus* of the surface, and it is the same g that occurs in the generalization of Euler's formula to surfaces.

Orientable surfaces

The genus turned out to be important for various deep issues in complex analysis, which in turn attracted attention to the topology of surfaces. It then turned out that there is a second class of surfaces, which differ from g-holed tori, but are closely related. The difference is that g-holed tori are orientable surfaces, which intuitively means that they have two distinct sides. They inherit this property from the complex plane, which has a top side and a bottom side because the spiral staircases are joined up in a way that preserves this distinction.

If instead you joined two flights of the staircase by twisting one floor upside down, the apparently separate sides would join together.

The possibility of this type of joining was first emphasized by Möbius, whose Möbius band has one side and one edge. Klein went one step further, by conceptually gluing a circular disc along the edge of the Möbius band to eliminate the edge altogether. The resulting surface, jokingly named the *Klein bottle*, has a single side and no edges at all. If we try to draw it sitting inside normal three-dimensional space, it has to pass through itself. But as an abstract surface in its own right (or as a surface sitting inside four-dimensional space) this self-intersection does not occur.

The theorem about g-holed tori can be restated like this: any orientable surface (of finite extent with no edges) is topologically equivalent to a sphere with g extra handles (where g might be zero). There is a similar classification of the non-orientable (one-sided) surfaces: they can be formed from a surface called the *projective plane* by adding g handles. The Klein bottle is a projective plane with one handle.

The combination of these two results is called the Classification Theorem for Surfaces. It tells us, up to topological equivalence, *all possible surfaces* (of finite extent with no edges). With the proof of this theorem, the topology of two-dimensional spaces – surfaces – could be considered known. That did not mean that every possible question about surfaces could be solved without further effort, but it did give a reasonable place to start from when considering more complicated issues. The Classification Theorem for Surfaces is a

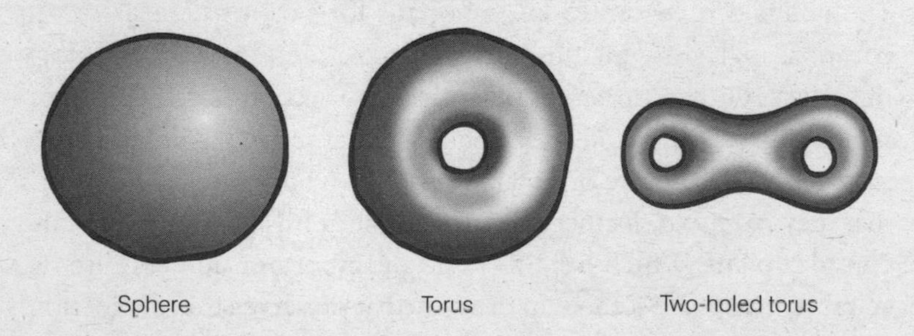

very powerful tool in two-dimensional topology.

When thinking about topology, it is often useful to suppose that the space concerned is everything that exists. There is no need to embed it in a surrounding space. So doing focuses attention on the *intrinsic* properties of the space. A vivid image is that of a tiny creature living on, say, a topological surface. How could such a creature, ignorant of any surrounding space, work out which surface it inhabits? How can we characterize such surfaces intrinsically?

By 1900 it was understood that one way to answer such questions is to consider closed loops in the surface, and how these loops can be deformed. For example, on a sphere any closed loop can be continuously deformed to a point – shrunk. For example, the circle running round the equator can be gradually moved towards the north pole, becoming smaller and smaller until it coincides with the north pole itself.

In contrast, every surface that is not equivalent to a sphere contains loops that cannot be deformed to points. Such loops pass through a hole, and the hole prevents them being shrunk. So the sphere can be characterized as *the* only surface in which any closed loop can be shrunk to a point.

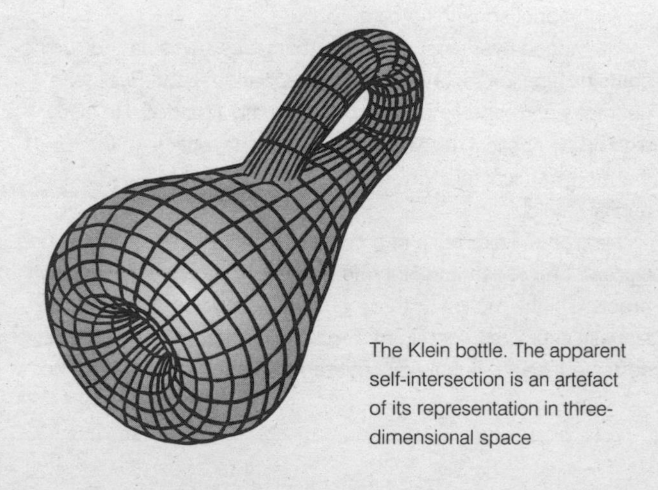

The Klein bottle. The apparent self-intersection is an artefact of its representation in three-dimensional space

Jules Henri Poincaré

1854–1912

Henri Poincaré was born in Nancy, France. His father Léon was Professor of Medicine at the University of Nancy, and his mother was Eugénie (née Launois). His cousin, Raymond Poincaré, became the French prime minister and was president of the French Republic during the First World War. Henri came top in every subject at school and was absolutely formidable in mathematics. He had an excellent memory and could visualize complicated shapes in three dimensions, which helped compensate for eyesight so poor that he could hardly see the blackboard, let alone what was written on it.

His first university job was in Caen in 1879, but by 1881 he had secured a far more prestigious job at the University of Paris. There he became one of the leading mathematicians of his age. He worked systematically – four hours per day in two two-hour periods, morning and late afternoon. But his thought processes were less organized, and he often started writing a research paper before he knew how it would end or where it would lead. He was highly intuitive, and his best ideas often arrived when he was thinking about something else.

He ranged over most of the mathematics of his day, including complex function theory, differential equations, non-Euclidean geometry and topology – which he virtually founded. He also worked on applications: electricity, elasticity, optics, thermodynamics, relativity, quantum theory, celestial mechanics and cosmology.

He won a major prize in a competition started in 1887 by King Oscar II of Sweden and Norway. The topic was the 'three-body problem' – the motion of three gravitating bodies. The work he actually submitted contained a major mistake, which he quickly corrected; as a result he discovered the possibility of what is now called chaos – irregular, unpredictable motion in a system governed by deterministic laws. He also wrote several bestselling popular science books: *Science and Hypothesis* (1901), *The Value of Science* (1905) and *Science and Method* (1908).

Topology in three dimensions

The natural next step after surfaces – two-dimensional topological spaces – is three dimensions. Now the objects of study are manifolds in Riemann's sense, except that notions of distance are ignored. In 1904 Henri Poincaré, one of the all-time greats in mathematics, was trying to understand three-dimensional manifolds. He introduced a number of techniques for achieving this goal. One of them, *homology*, studies the relations between regions in the manifold and their boundaries. Another, *homotopy*, looks at what happens to closed loops in the manifold as the loops are deformed.

Homotopy is closely related to the methods that had served so well for surfaces, and Poincaré looked for analogous results in three dimensions. Here he was led to one of the most famous questions in the whole of mathematics.

He knew the characterization of the sphere as the only surface in which any closed loop can be shrunk. Did a similar characterization work in three dimensions? For a time he assumed that it did; in fact, this seemed so obvious that he didn't even notice he was making an assumption. Later he realized that one plausible version of this statement is actually wrong, while another closely related formulation seemed difficult to prove but might well be true. He posed a question, later reinterpreted as the Poincaré conjecture – if a three-dimensional manifold (without boundary, of finite extent, and so on) has the property that any loop in it can be shrunk to a point, then that manifold must be topologically equivalent to the 3-sphere (a natural three-dimensional analogue of a sphere).

Subsequent attempts to prove the conjecture succeeded for generalizations to four or more dimensions. Topologists continued to struggle with the original Poincaré conjecture, in three dimensions, without success.

In the 1980s William Thurston came up with an idea that might make it possible to sneak up on the Poincaré conjecture by being more ambitious. His geometrization conjecture goes further, and applies to all three-dimensional manifolds, not just those in which every loop can be shrunk. Its starting-point is an interpretation of the classification of surfaces in terms of non-Euclidean geometry.

The torus can be obtained by taking a square in the Euclidean plane, and identifying opposite edges. As such, it is flat – has zero curvature. The sphere has constant positive curvature. A torus with two or more holes can be represented as a surface of constant negative curvature. So the topology of surfaces can be reinterpreted in terms of three types of geometry, one Euclidean and two non-Euclidean, namely, Euclidean geometry itself, elliptic geometry (positive curvature) and hyperbolic geometry (negative curvature).

Might something similar hold in three dimensions? Thurston pointed out some complications: there are eight types of geometry to consider, not three. And it is no longer possible to use a single geometry for a given manifold: instead, the manifold must be cut into several pieces, using one geometry for each. He formulated his geometrization conjecture: there is always a systematic way to cut up a three-dimensional manifold into pieces, each corresponding to one of the eight geometries.

The Poincaré conjecture would be an immediate consequence, because the condition that all loops shrink rules out seven geometries, leaving just the geometry of constant positive curvature – that of the 3-sphere.

An alternative approach emerged from Riemannian geometry. In 1982 Richard Hamilton introduced a new technique into the area, based on the mathematical ideas used by Albert Einstein in general relativity. According to Einstein, space-time can be considered as curved, and the curvature describes the force of gravity. Curvature is measured by the so-called curvature tensor, and this has a simpler relative known as the Ricci tensor after its inventor Gregorio Ricci-Curbastro. Changes in the geometry of the universe over time are governed by the Einstein equations, which say that the stress tensor is proportional to the curvature. In effect, the gravitational bending of the universe tries to smooth itself out as time passes, and the Einstein equations quantify that idea.

The same game can be played using the Ricci version of curvature, and it leads to the same kind of behaviour: a surface that obeys the equations for the Ricci flow will naturally tend to simplify its own geometry by redistributing its curvature more equitably. Hamilton showed that the two-dimensional Poincaré conjecture can be proved using the Ricci flow – basically, a surface in which all loops shrink simplifies itself so much as it follows the Ricci flow that it ends up as a perfect sphere. Hamilton also suggested generalizing this approach to three dimensions, and made some progress along those lines, but hit some difficult obstacles.

Perelman

In 2002 Grigori Perelman caused a sensation by placing several papers on the arXiv, a website for physics and mathematics research that lets researchers provide public access to unrefereed, often ongoing, work. The aim of the website is to avoid long delays that occur while papers are being refereed for official publication. Previously, this role had been played by informal preprints. Ostensibly Perelman's papers were about the Ricci flow, but it became clear that if the work was correct, it would imply the geometrization conjecture, hence that of Poincaré.

The basic idea is the one suggested by Hamilton. Start with an arbitrary three-dimensional manifold, equip it with a notion of

What topology did for them

One of the simplest topological invariants was invented by Gauss. In a study of electrical and magnetic fields, he became interested in how two closed loops can link together. He invented the linking number, which measures how many times one loop winds round the other one. If the linking number is non-zero, then the loops cannot be separated by a topological transformation. However, this invariant does not completely solve the problem of determining when two linked loops cannot be separated, because sometimes the linking invariant is zero but the links cannot be separated.

He even developed an analytic formula for this number, by integrating a suitable quantity along the curve concerned. Gauss's discoveries provided a foretaste of what is now a huge area of mathematics, algebraic topology.

(Below) Loops with linking number 3;
(Right) These links cannot be separated topologically, even though they have linking number 0

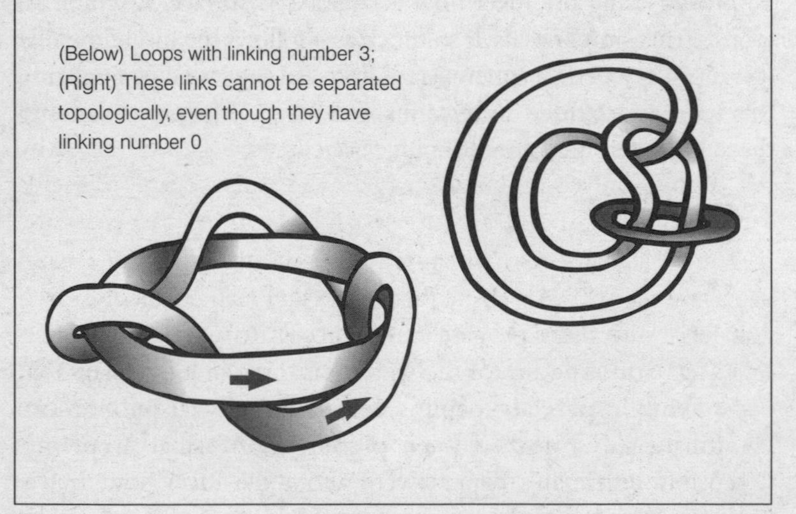

distance so that the Ricci flow makes sense, and let the manifold follow the flow, simplifying itself. The main complication is that singularities can develop, where the manifold pinches together and ceases to be smooth. At singularities, the proposed method breaks down. The new idea is to cut the manifold apart near such a singularity, cap off the resulting holes and let the flow continue. If the manifold manages to simplify itself completely after only finitely many singularities have arisen, each piece will support just one of the eight geometries, and reversing the cutting operations (surgery) tells us how to glue those pieces back together to reconstruct the manifold.

The Poincaré conjecture is famous for another reason: it is one of the eight Millennium Mathematics Problems selected by the Clay Institute, and as such its solution – suitably verified – attracts a million-dollar prize. However, Perelman had his own reasons for not wanting the prize – indeed, any reward save the solution itself – and therefore had no strong reason to expand his often-cryptic papers on the arXiv into something more suitable for publication.

Experts in the area therefore developed their own versions of his ideas, trying to fill in any apparent gaps in the logic, and generally tidy up the work into something acceptable as a genuine proof. Several such attempts were published, and a comprehensive and definitive version of Perelman's proof has now been accepted by the topological community. In 2006 he was awarded a Fields Medal for his work in this area, which he declined. Not everyone craves worldly success.

Topology and the real world

Topology was invented because mathematics could not function without it, stimulated by a number of basic questions in areas like complex analysis. It tackles the question 'what shape is this thing?' in a very simple but deep form. More conventional geometric concepts, such as lengths, can be seen as adding extra detail to the basic information captured by topology.

A few early forerunners of topology exist, but it did not really become a branch of mathematics with its own identity and power

Grigori Perelman
1966 –

Perelman was born in what was then the Soviet Union in 1966. As a student he was a member of the USSR team competing in the International Mathematical Olympiad, and won a gold medal with a 100% score. He has worked in the USA and at the Steklov Institute in St Petersburg, but currently holds no specific academic position. His increasingly reclusive nature has added an unusual human dimension to the mathematical story. It is perhaps a pity that this story reinforces the stereotype of the eccentric mathematician.

until the middle of the 19th century, when mathematicians obtained a fairly complete understanding of the topology of surfaces, two-dimensional shapes. The extension to higher dimensions was given a huge boost in the late 19th and early 20th centuries, notably with the investigations of Henri Poincaré. Further advances occurred in the 1920s; the subject really took off in the 1960s, although, ironically, it pretty much lost contact with applied science.

Confounding those critical of the abstraction of 20th century pure mathematics, the resulting theory is now vital to several areas of mathematical physics. Even its most intractable obstacle, the Poincaré conjecture, has now been overcome. In retrospect, the main difficulties in developing topology were internal ones, best solved by abstract means; connections with the real world had to wait until the techniques were sorted out properly.

What topology does for us

In 1956 James Watson and Francis Crick discovered the secret of life the double-helix structure of the DNA molecule, the backbone on which genetic information is stored and manipulated. Today the topology of knots is being used to understand how the two strands of the helix disentangle as the genetic blueprint controls the development of a living creature.

The DNA helix is like a two-stranded rope, with each strand twisted repeatedly around the other. When a cell divides, the genetic information is transferred to the new cells by splitting the strands apart, copying them and joining the new strands to the old in pairs. Anyone who has tried to separate the strands of a long piece of rope knows how tricky this process is: the strands tangle in lumps as you try to pull them apart. In fact DNA is much worse: the helices themselves are supercoiled as if the rope itself has been wound into a coil. Imagine several kilometres of fine thread stuffed into a tennis-ball and you have some idea of how tangled the DNA in a cell must be.

The genetic biochemistry must ravel and unravel this tangled thread, rapidly, repeatedly and faultlessly; the very chain of life depends upon it. How? Biologists tackle the problem by using enzymes to break the DNA chain into pieces, small enough to investigate in detail. A segment of DNA is a complicated molecular knot, and the same knot can look very different after a few tucks and turns have distorted its appearance.

The new techniques for studying knots open up new lines of attack on molecular genetics. No longer a plaything of the pure mathematicians, the topology of knots is becoming an important practical issue in biology. A recent discovery is a mathematical connection between the amount of twisting in the DNA helix and the amount of supercoiling.

CHAPTER 16

The Fourth Dimension

Geometry out of this world

In his science fiction novel *The Time Machine*, Herbert George Wells described the underlying nature of space and time in a way that we now find familiar, but which must have raised some eyebrows among his Victorian readers: 'There are really four dimensions, three which we call the three planes of Space, and a fourth, Time'. To set up the background for his story, he added: 'There is, however, a tendency to draw an unreal distinction between the former three dimensions and the latter, because it happens that our consciousness moves intermittently in one direction along the latter from the beginning to the end of our lives. But some philosophical people have been asking why three dimensions particularly – why not another direction at right angles to the three? – and have even tried to construct a four-dimensional geometry'. His protagonist then goes one better, overcomes the alleged limitations of human consciousness and travels along the fourth dimension of time, as if it were a normal dimension of space.

The fourth dimension

The art of the science fiction writer is suspension of disbelief, and Wells achieved this by informing his readers that 'Professor Simon Newcomb was expounding this to the New York Mathematical Society only a month or so ago'. Here Wells was probably referring to a real event; we know that Newcomb, a prominent astronomer, gave a lecture on four-dimensional space at roughly the right time. His lecture reflected a major change in mathematical and scientific thinking, freeing these subjects from the traditional assumption that space must always have three dimensions. This does not imply that time travel is possible, but it gave Wells an excuse to make penetrating observations about present-day human nature by displacing his time traveller into a disturbing future.

The Time Machine, published in 1895, resonated with a Victorian obsession with the fourth dimension, in which an additional, unseen dimension of space was invoked as a place for ghosts, spirits, or even God to reside. The fourth dimension was championed by charlatans, exploited by novelists, speculated upon by scientists and formalized by mathematicians. Within a few decades, not only was four-dimensional space standard in mathematics: so were spaces with any number of dimensions – five, ten, a billion, even infinity. The techniques and thought-patterns of multidimensional geometry were being used routinely in every branch of science – even biology and economics.

Higher-dimensional spaces remain almost unknown outside the scientific community, but very few areas of human thought could now function effectively without these techniques, remote though they may seem from ordinary human affairs. Scientists trying to unify the two great theories of the physical universe, relativity and quantum mechanics, are speculating that space may actually have nine dimensions, or ten, rather than the three that we normally perceive. In a rerun of the fuss about non-Euclidean geometry, space of three dimensions is increasingly being viewed as just one possibility out of many, rather than the only kind of space that is possible.

These changes have come about because terms like space and dimension are now interpreted in a more general manner, which agrees with the usual dictionary meanings in the familiar contexts of a TV screen or our normal surroundings, but opens up new possibilities. To mathematicians, a *space* is a collection of objects together with some notion of the distance between any two of those objects. Taking a hint from Descartes's idea of coordinates, we can define the dimension of such a space to be *how many numbers* are required to specify an object. With points as the objects, and the usual notion of distance in the plane or space, we find that the plane has two dimensions and space has three. However, other collections of objects may have four dimensions, or more, depending on what the objects are.

For example, suppose that the objects are spheres in three-dimensional space. It takes four numbers (x, y, z, r) to specify a sphere: three coordinates (x, y, z) for its centre, plus the radius r. So the space of all spheres in ordinary space has four dimensions. Examples like this show that natural mathematical questions can easily lead to higher-dimensional spaces.

Indeed, modern mathematics goes further. Abstractly, space of four dimensions is *defined* as the set of all quadruples (x_1, x_2, x_3, x_4) of numbers. More generally, space of n dimensions – for any whole number n – is defined as the set of all n-tuples $(x_1, x_2, \ldots, x_n)$ of numbers. In a sense, that is the whole story; the intriguing and baffling notion of many dimensions collapses to a triviality: long lists of numbers.

That viewpoint is now clear, but historically, it took a long time to become established. Mathematicians argued, often very forcibly, about the meaning and reality of higher-dimensional spaces. It took about a century for the ideas to become widely accepted. But the applications of such spaces, and the geometric imagery that went with them, proved so useful that the underlying mathematical issues ceased to be controversial.

Three- or four-dimensional space

Ironically, today's conception of higher-dimensional spaces emerged from algebra, not geometry, as a consequence of a failed attempt to develop a three-dimensional number system, analogous to the two-dimensional system of complex numbers. The distinction between two and three dimensions goes back to Euclid's *Elements*. The first part of the book is about the geometry of the plane, a space of two dimensions. The second part is about solid geometry – the geometry of three-dimensional space. Until the 19th century, the word dimension was limited to these familiar contexts.

Greek geometry was a formalization of the human senses of sight and touch, which allow our brains to build internal models of positional relationships of the outside world. It was constrained by the limitations of our own senses, and of the world in which we live. The Greeks thought that geometry described the *real* space in which we live, and they assumed that physical space has to be Euclidean. The mathematical question 'can four-dimensional space exist in some conceptual sense?' became confused with the physical question 'can a *real* space with four dimensions exist?' And that question was further confused with 'can there be four dimensions *within our own familiar space?*' to which the answer is 'no'. So it was generally believed that four-dimensional space is impossible.

Geometry began to free itself from this restricted viewpoint when the algebraists of Renaissance Italy unwittingly stumbled upon a profound extension of the number concept, by accepting the existence of a square root of minus one. Wallis, Wessel, Argand and Gauss worked out how to interpret the resulting complex numbers as points in a plane, freeing numbers from the one-dimensional shackles of the real number line. In 1837, The Irish mathematician William Rowan Hamilton reduced the whole topic to algebra, by

William Rowan Hamilton
1805–1865

Hamilton was so precocious mathematically that he was made Professor of Astronomy at Trinity College Dublin while still an undergraduate, at the age of 21. This appointment made him Royal Astronomer of Ireland.

He made numerous contributions to mathematics, but the one that he himself believed to be most significant was the invention of quaternions. He tells us that 'Quaternions ... started into life, fully grown, on the 16th of October, 1843, as I was walking with Lady Hamilton to Dublin, and came up to Brougham Bridge. That is to say, I then and there felt the galvanic circuit of thought closed, and the sparks which fell from it were the fundamental equations between *i*, *j*, *k*; *exactly such* as I have used them ever since. I pulled out, on the spot, a pocketbook, which still exists, and made an entry, on which, *at the very moment*, I felt that it might be worth my while to expend the labour of at least ten (or it might be fifteen) years to come. I felt a *problem* to have been at that moment *solved*, an intellectual *want relieved*, which had haunted me for at least *fifteen years* before.'

Hamilton immediately carved the equation

$$i^2 = j^2 = k^2 = ijk = -1$$

in the stone of the bridge.

defining a complex number $x + iy$ to be a *pair* of real numbers (x, y). He further defined addition and multiplication of pairs by the rules

$$(x, y) + (u, v) = (x + u, y + v)$$

$$(x, y)(u, v) = (xu - yv, xv + yu)$$

In this approach, a pair of the form $(x,0)$ behaves just like the real number x, and the special pair (0,1) behaves like i. The idea is simple, but appreciating it requires a sophisticated concept of mathematical existence.

Hamilton then set his sights on something more ambitious. It was well known that complex numbers make it possible to solve many problems about the mathematical physics of systems in the plane, using simple and elegant methods. A similar trick for three-dimensional space would be invaluable. So he tried to invent a *three-dimensional* number system, in the hope that the associated calculus would solve important problems of mathematical physics in three-dimensional space. He tacitly assumed that this system would satisfy all the usual laws of algebra. But despite heroic efforts, he could not find such a system.

Eventually, he discovered why. It's impossible.

Among the usual laws of algebra is the *commutative law of multiplication*, which states that $ab = ba$. Hamilton had been struggling for years to devise an effective algebra for three dimensions. Eventually he found one, a number system that he called *quaternions*. But it was an algebra of four dimensions, not three, and its multiplication was not commutative.

Quaternions resemble complex numbers, but instead of one new number, i, there are three: i, j, k. A quaternion is a combination of these, for example 7 + 8i – 2j + 4k. Just as the complex numbers are two-dimensional, built from two independent quantities 1 and i, so the quaternions are *four*-dimensional, built from four *independent* quantities 1, i, j and k. They can be formalized algebraically as quadruples of real numbers, with particular rules for addition and multiplication.

Higher-dimensional space

When Hamilton made his breakthrough, mathematicians were already aware that spaces of high dimension arise entirely naturally, and have sensible physical interpretations, when the basic elements of space are something other than points. In 1846 Julius Plücker pointed out that it takes *four* numbers to specify a *line* in space. Two of those numbers determine where the line hits some fixed plane; two more determine its direction relative to that plane. So, considered as a collection of *lines*, our familiar space already has four dimensions, not three. However, there was a vague feeling that this construction was rather artificial, and that spaces made from four dimensions worth of points were unnatural. Hamilton's quaternions had a natural interpretation as rotations, and their algebra was compelling. They were as natural as complex numbers – so four-dimensional space was as natural as a plane.

The idea quickly went beyond just four dimensions. While Hamilton was promoting his beloved quaternions, a mathematics teacher named Hermann Günther Grassmann was discovering an extension of the number system to spaces with any number of dimensions. He published his idea in 1844 as *Lectures on Lineal Extension*. His presentation was mystical and rather abstract, so the work attracted little attention. In 1862, to combat the lack of interest, he issued a revised version, often translated as *The Calculus of Extension*, which was intended to be more comprehensible. Unfortunately, it wasn't.

Despite its cool reception, Grassmann's work was of fundamental importance. He realized that it was possible to replace the four units, 1, i, j and k, of quaternions by any number of units. He called combinations of these units *hypernumbers*. He understood that his approach had limitations. You had to be careful not to expect too much from the arithmetic of hypernumbers; slavishly following the traditional laws of algebra seldom led anywhere.

Meanwhile, physicists were developing their own notions of higher-dimensional spaces, motivated not by geometry, but by Maxwell's equations for electromagnetism. Here both the electric and magnetic fields are *vectors* – having a *direction* in three-dimensional space as well as a magnitude. Vectors are arrows, if you wish, aligned with the electric or magnetic field. The length of the arrow shows how strong the field is, and its direction shows which way the field is pointing.

In the notation of the time, Maxwell's equations were eight in number, but they included two groups of three equations, one for each component of the electric or the magnetic field in each of the three directions of space. It would make life much easier to devise a formalism that collected each such triple into a single vector equation. Maxwell achieved this using quaternions, but his approach was clumsy. Independently, the physicist Josiah Willard Gibbs and the engineer Oliver Heaviside found a simpler way to represent vectors algebraically. In 1881 Gibbs printed a private pamphlet, *Elements of Vector Analysis*, to help his students. He explained that his ideas had been developed for convenient use rather than mathematical elegance. His notes were written up by Edwin Wilson, and they published a joint book *Vector Analysis* in 1901. Heaviside came up with the same general ideas in the first volume of his *Electromagnetic Theory* in 1893 (the other two volumes appeared in 1899 and 1912).

The different systems of Hamilton's quaternions, Grassmann's hypercomplex numbers and Gibbs's vectors rapidly converged to the same mathematical description of a vector: it is a triple (x, y, z) of numbers. After 250 years, the world's mathematicians and physicists had worked their way right back to Descartes – but now the coordinate notation was only part of the story. Triples did not just represent points: they represented directed magnitudes. It made a huge difference – not to the formalism, but to its interpretation, its physical *meaning*.

Mathematicians wondered just how many hypercomplex number systems there might be. To them, the question was not 'are they useful?' but 'are they interesting?' So, mathematicians mainly focused on the algebraic properties of systems of n-hypercomplex numbers, for any n. These were, in fact, n-dimensional spaces, *plus* algebraic operations, but to begin with everyone thought algebraically and the geometric aspects were played down.

Differential geometry

Geometers responded to the algebraists' invasion of their territory by reinterpreting hypercomplex numbers geometrically. The key figure here was Riemann. He was working for his 'Habilitation' which gave him the right to charge lecture fees to students. Candidates for Habilitation must give a special lecture on their own research. Following the usual procedure, Gauss asked Riemann to propose a number of topics, from which Gauss would make the final choice. One of Riemann's proposals was On the Hypotheses Which Lie at the Foundation of Geometry, and Gauss, who had been thinking about the same question, chose that topic.

Riemann was terrified – he disliked public speaking and he hadn't fully worked out his ideas. But what he had in mind was explosive: a geometry of n dimensions, by which he meant a system of n coordinates $(x_1, x_2, ..., x_n)$, equipped with a notion of distance between nearby points. He called such a space a manifold. This proposal was radical enough, but there was another, even more radical feature: manifolds could be curved. Gauss had been studying the curvature of surfaces, and had obtained a beautiful formula which represented curvature intrinsically – that is, in terms of the surface alone, not of the space in which it was embedded.

Riemann had intended to develop a similar formula for the curvature of a manifold, generalizing Gauss's formula to n dimensions. This formula would also be intrinsic to the manifold – it would not make explicit use of any containing space. Riemann's efforts to develop the notion of curvature in a space of n dimensions led him to the brink of a nervous breakdown. What made matters worse was that at the same time, he was helping Gauss's colleague Weber, who was trying to understand electricity. Riemann battled on, and the interplay between electrical and magnetic forces led him to a new concept of force based on geometry. He had the same insight that led Einstein to general relativity, decades later: forces can be replaced by the curvature of space.

In traditional mechanics, bodies travel along straight lines unless diverted by a force. In curved geometries, straight lines need not exist and paths are curved. If space is curved, what you experience when you are obliged to deviate from a straight line feels like a force. Now Riemann had the insight he needed to develop his lecture, which he gave in 1854. It was a major triumph. The ideas quickly spread, to growing excitement. Soon scientists were giving popular lectures on the new geometry. Among them was Hermann von Helmholtz, who gave talks about beings that lived on a sphere or some other curved surface.

The technical aspects of Riemann's geometry of manifolds, now called differential geometry, were further developed by Eugenio Beltrami, Elwin Bruno Christoffel and the Italian school under Gregorio Ricci and Tullio Levi-Civita. Later, their work turned out to be just what Einstein needed for general relativity.

Matrix algebra

Algebraists had also been busy, developing computational techniques for n-variable algebra – the formal symbolism of n-dimensional space. One of these techniques was the algebra of matrices, rectangular arrays of numbers, introduced by Cayley in 1855. This formalism arose naturally from the idea of a change of coordinates. It had become commonplace to simplify algebraic formulas by replacing variables such as x and y by their linear combinations, for example

$$u = ax + by$$

$$v = cx + dy$$

for constants a, b, c and d. Cayley represented the pair (x, y) as a column vector and the coefficients by a 2×2 table, or *matrix*. With a suitable definition of multiplication, he could rewrite the coordinate change as

$$\begin{bmatrix} u \\ v \end{bmatrix} = \begin{bmatrix} a & b \\ c & d \end{bmatrix} \begin{bmatrix} x \\ y \end{bmatrix}$$

The method extended easily to tables with any number of rows and columns, representing linear changes in any number of coordinates.

Matrix algebra made it possible to calculate in n-dimensional space. As the new ideas sank in, a geometric language for n-dimensional space came into being, supported by a formal algebraic computational system. Cayley thought that his idea was no more than a notational convenience, and predicted that it would never acquire any applications. Today, it is indispensable throughout science, especially in areas like statistics. Medical trials are a great consumer of matrices, which are used to work out which associations between cause and effect are statistically significant.

The geometric imagery made it easier to prove theorems. Critics countered that these newfangled geometries referred to spaces that didn't exist. The algebraists fought back by pointing out that the algebra of n variables most certainly did exist, and anything that helped advance many different areas of mathematics must surely be interesting. George Salmon wrote: 'I have already completely discussed this problem [solving a certain system of equations] when we are given three equations in three variables. The question now before us may be stated as the corresponding problem in space of p dimensions. But we consider it as a purely algebraic question, apart from any geometrical considerations. We shall however retain a little of geometrical *language*... because we can thus more readily see how to apply to a system of p equations, processes analogous to those which we have employed in a system of three.'

What high-dimensional geometry did for them

Around 1907 the German mathematician Hermann Minkowski formulated Einstein's theory of special relativity in terms of a four-dimensional *space-time*, combining one-dimensional time and three-dimensional space into a single mathematical object. This is known as *Minkowski space-time*.

The requirements of relativity imply that the natural metric on Minkowski space-time is not the one determined by Pythagoras's theorem, in which the square of the distance from a point (x, t) to the origin is $x^2 + t^2$. Instead, this expression should be replaced by the interval $x^2 - c^2t^2$, where c is the speed of light. The crucial change here is the minus sign, which implies that events in space-time are associated with two cones. One cone (here a triangle because space has been reduced to one dimension) represents the future of the event, the other its past. This geometric representation is employed almost universally by modern physicists.

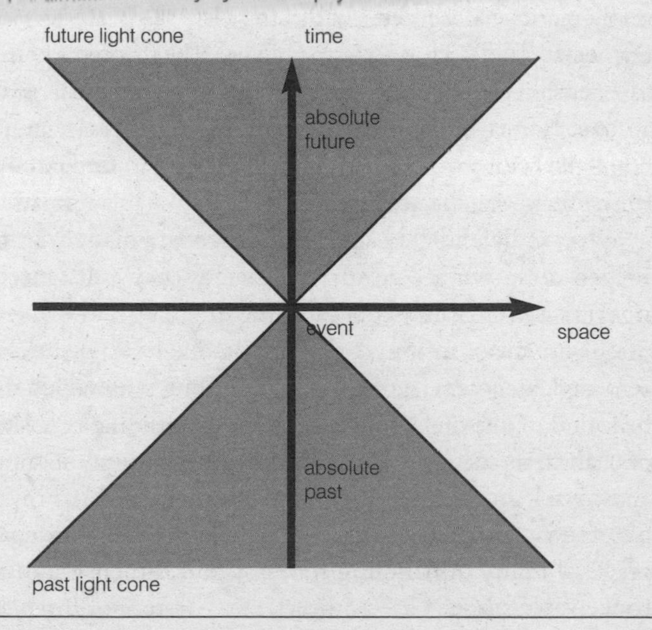

Real space

Do higher dimensions exist? Of course, the answer depends on what we mean by 'exist', but people tend not to understand that kind of thing, especially when their emotions are aroused. The issue came to a head in 1869. In a famous address to the British Association, later reprinted as *A Plea for the Mathematician*, James Joseph Sylvester pointed out that generalization is an important way to advance mathematics. What matters, said Sylvester, is what is *conceivable*, not what corresponds directly to physical experience. He added that with a little practice it is perfectly possible to visualize four dimensions, so four-dimensional space is conceivable.

This so infuriated the Shakespearean scholar Clement Ingleby that he invoked the great philosopher Immanuel Kant to prove that three-dimensionality is an essential feature of space, completely missing Sylvester's point. The nature of real space is irrelevant to the mathematical issues. Nevertheless, for a time most British mathematicians agreed with Ingleby. But some continental mathematicians did not. Grassmann said: 'The theorems of the Calculus of Extension are not merely translations of geometrical results into an abstract language; they have a much more general significance, for while the ordinary geometry remains bound to three dimensions of [physical] space, the abstract science is free of this limitation'.

Sylvester defended his position: 'There are many who regard the alleged notion of a generalized space as only a disguised form of algebraic formulization; but the same might be said with equal truth of our notion of infinity, or of impossible lines, or lines making a zero angle in geometry, the utility of dealing with which no one will be found to dispute. Dr Salmon in his extension of Chasles's theory of characteristics to surfaces, Mr Clifford in a question of probability, and myself in the theory of partitions, and also in my paper on barycentric projection, have all felt and given evidence on the practical utility of handling space of four dimensions as if it were conceivable space'.

Multidimensional space

In the end, Sylvester won the debate. Nowadays mathematicians consider something to exist if it is not logically contradictory. It may contradict physical experience, but that is irrelevant to *mathematical* existence. In this sense, multidimensional spaces are just as real as the familiar space of three dimensions, because it is just as easy to provide a formal definition.

The mathematics of multidimensional spaces, as now conceived, is purely algebraic, and based on obvious generalizations from low-dimensional spaces. For example, every point in the plane (a two-dimensional space) can be specified by its two coordinates, and every point in three-dimensional space can be specified by its three coordinates. It is a short step to define a point in four-dimensional space as a set of four coordinates, and more generally to define a point in n-dimensional space as a list of n coordinates. Then n-dimensional space itself (or n-space for short) is just the set of all such points.

Similar algebraic machinations let you work out the distance between any two points in n-space, the angle between any two lines, and so on. From there on out, it's a matter of imagination: most sensible geometric shapes in two or three dimensions have straightforward analogues in n dimensions, and the way to find them is to describe the familiar shapes using the algebra of coordinates and then extend that description to n coordinates.

For example, a circle in the plane, or a sphere in 3-space, consists of all points that lie at a fixed distance (the radius) from a chosen point (the centre). The obvious analogue in n-space is to consider all points that lie at a fixed distance from a chosen point. Using the formula for distances, this becomes a purely algebraic condition, and the resulting object is known as an $(n - 1)$-dimensional hypersphere, or $(n - 1)$-sphere for short. The dimension drops from n to $n - 1$ because, for example, a circle in 2-space is a curve, which is a one-dimensional object; similarly a sphere in space is a two-

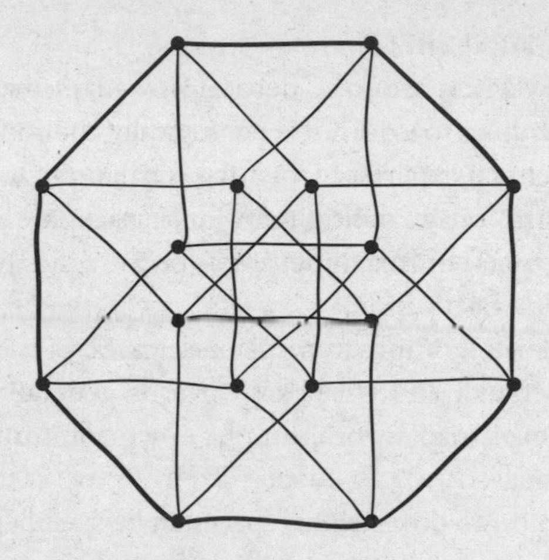

A four-dimensional hypercube, projected on to the plane

dimensional surface. A *solid* hypersphere in *n* dimensions is called an *n*-ball. So the Earth is a 3-ball and its surface is a 2-sphere.

Nowadays, this point of view is called *linear algebra*. It is used throughout mathematics and science, especially in engineering and statistics. It is also a standard technique in economics. Cayley stated that his matrices were unlikely ever to have any practical application. He could not have been more wrong.

By 1900 Sylvester's predictions were coming true, with an explosion of mathematical and physical areas where the concept of multidimensional space was having a serious impact. One such area was Einstein's relativity, best considered as a special kind of four-dimensional space-time geometry. In 1908 Hermann Minkowski realized that the three coordinates of ordinary space, together with an extra one for time, form a four-dimensional *space-time*. Any point in space-time is called an *event*: it is like a point particle that winks into existence at just one moment in time, and then winks out again. Relativity is really about the physics of events. In traditional mechanics, a particle moving through space occupies coordinates

$(x(t), y(t), z(t))$ at time t, and this position changes as time passes. From Minkowski's space-time viewpoint, the collection of all such points is a curve in space-time, the *world line* of the particle, and it is a single object in its own right, existing for all time. In relativity, the fourth dimension has a single, fixed interpretation: *time*.

The subsequent incorporation of gravity, achieved in general relativity, made heavy use of Riemann's revolutionary geometries, but modified to suit Minkowski's representation of the geometry of flat space-time – that is, what space and time do when no mass is present to cause gravitational distortions, which Einstein modelled as curvature.

Mathematicians preferred a more flexible notion of dimensionality and space and as the late 19th century flowed into the early 20th, mathematics itself seemed, ever more, to demand acceptance of multidimensional geometry. The theory of functions of two complex variables, a natural extension of complex analysis, required thinking about space of two complex dimensions – but each complex dimension boils down to two real ones, so like it or not you are looking at a four-dimensional space. Riemann's manifolds and the algebra of many variables provided further motivation.

Generalized coordinates

Yet another stimulus towards multidimensional geometry was Hamilton's 1835 reformulation of mechanics in terms of generalized coordinates, a development initiated by Lagrange in his *Analytical Mechanics* of 1788. A mechanical system has as many of these coordinates as it has degrees of freedom – that is, ways to change its state. In fact the number of degrees of freedom is just dimension in disguise.

For example, it takes six generalized coordinates to specify the configuration of a rudimentary bicycle: one for the angle at which the handlebars sit relative to the frame, one each for the angular

What high-dimensional geometry does for us

Your mobile phone makes essential use of multidimensional spaces. So do your Internet connection, your satellite or cable TV and virtually any other piece of technology that sends or receives messages. Modern communications are digital. All messages, even telephone voice messages, are converted into patterns of 0s and 1s – binary numbers.

Communications are not much use unless they are reliable – the message that is received should be exactly the same as the one that was sent. Electronic hardware cannot guarantee this kind of accuracy, because interference, or even a passing cosmic ray, can cause mistakes. So, electronic engineers use mathematical techniques to put signals into code, in such a way that errors can be detected, and even corrected. The basis of these codes is the mathematics of multidimensional spaces.

Such spaces turn up because a string of, say, ten binary digits, or *bits*, such as 1001011100, can profitably be viewed as a point in a ten-dimensional space with coordinates restricted either to 0 or to 1. Many important questions about error-detecting and error-correcting codes are best tackled in terms of the geometry of this space.

For example, we can detect (but not correct) a single error if we code every message by replacing every 0 by 00 and every 1 by 11. Then, a message such as 110100 codes as 111100110000. If this is received as 111000110000, with an error in the fourth bit, we know something has

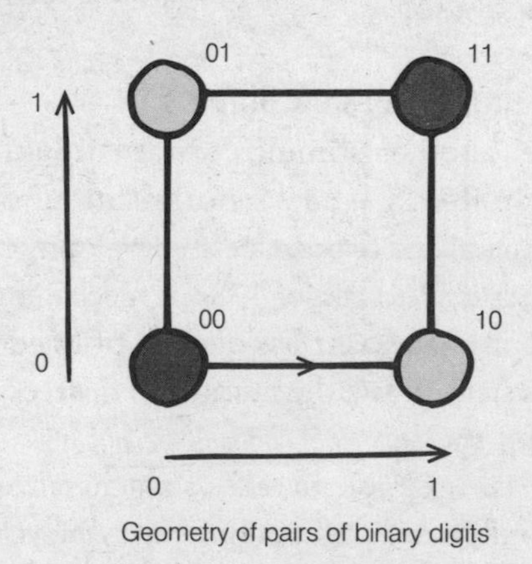

Geometry of pairs of binary digits

gone wrong because the boldface pair 10 should not occur. But we don't know whether it should have been 00 or 11. This can be neatly illustrated in a two-dimensional figure (corresponding to the length 2 of the code words 00 and 11). By thinking of the bits in the code words as coordinates relative to two axes (corresponding to the first and second digits of the code word, respectively) we can draw a picture in which the valid code words 00 and 11 are diagonally opposite corners of a square.

Any single error changes them to code words at the other two corners – which are not valid code words. However, because these corners are adjacent to both of the valid code words, different errors can lead to the same result. To get an error-correcting code, we can use code words of length three and encode 0 as 000 and 1 as 111. Now the code words live at the corners of a cube in three-dimensional space. Any single error results in an adjacent code word; moreover, each such invalid code word is adjacent to only one of the valid code words 000 or 111.

This approach to coding of digital messages was pioneered by Richard Hamming in 1947. The geometric interpretation came soon after, and it has proved crucial for the development of more efficient codes.

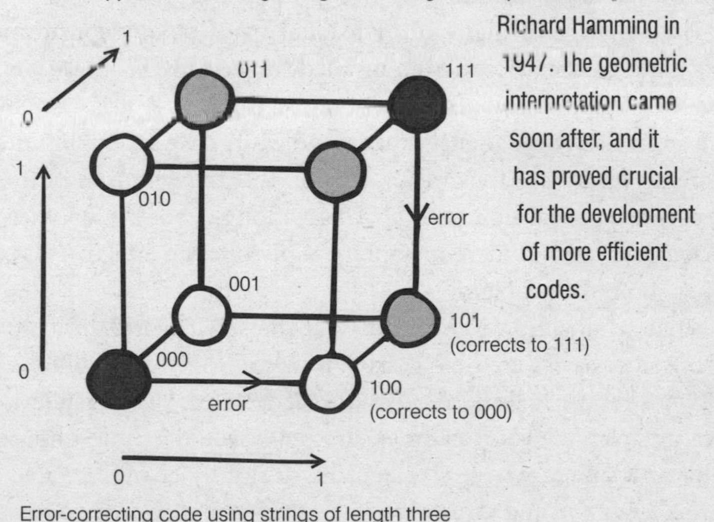

Error-correcting code using strings of length three

positions of the two wheels, another for the axle of the pedals, two more for the rotational positions of the pedals themselves. A bicycle is, of course, a three-dimensional *object* – but the space of possible *configurations* of the bicycle is six-dimensional, which is one of the reasons why learning to ride a bicycle is hard until you get the knack. Your brain has to construct an internal representation of how those six variables interact – you have to learn to navigate in the six-dimensional geometry of bicycle-space. For a moving bicycle, there are six corresponding velocities to worry about too: the dynamics is, in essence, *twelve*-dimensional.

By 1920 this concurrence of physics, mathematics and mechanics had triumphed, and the use of geometric language for many-variable problems – multidimensional geometry – had ceased to raise eyebrows, except perhaps among philosophers. By 1950, the process had gone so far that mathematicians' natural tendency was to formulate everything in n dimensions from the beginning. Limiting theories to two or three dimensions seemed old-fashioned and ridiculously confining.

The language of higher-dimensional space rapidly spread into every area of science, and even invaded subjects like economics and genetics. Today's virologists, for instance, think of viruses as points in a space of DNA sequences that could easily have several hundred dimensions. By this they mean, at root, that the genomes of these viruses are several hundred DNA bases long – but the geometric image goes beyond mere metaphor: it provides an effective way to think about the problem.

None of this, however, means that the spirit world exists, that ghosts now have a credible home, or that one day we might (as in Edwin Abbott's *Flatland*) receive a visit from the Hypersphere, a creature from the Fourth Dimension, who would manifest himself to us as a sphere whose size kept mysteriously changing, able to shrink to a point and vanish from our universe. However, physicists working in the theory of superstrings currently think that our universe may actually have *ten* dimensions, not four. Right now they think that we've never noticed the extra six dimensions because they are curled up too tightly for us to detect them.

Multidimensional geometry is one of the most dramatic areas in which mathematics appears to lose all touch with reality. Since physical space is three-dimensional, how can spaces of four or more dimensions exist? And even if they can be defined mathematically, how can they possibly be useful?

The mistake here is to expect mathematics to be an obvious, literal translation of reality, observed in the most direct manner. We are in fact surrounded by objects that can best be described by a large number of variables, the 'degrees of freedom' of those objects. To state the position of a human skeleton requires at least 100 variables, for example. Mathematically, the natural description of such objects is in terms of high-dimensional spaces, with one dimension for each variable.

It took mathematicians a long time to formalize such descriptions, and even longer to convince anyone else that they are useful. Today, they have become so deeply embedded in scientific thinking that their use has become a reflex action. They are standard in economics, biology, physics, engineering, astronomy... the list is endless.

The advantage of high-dimensional geometry is that it brings human visual abilities to bear on problems that are not initially visual at all. Because our brains are adept at visual thinking, this formulation can often lead to unexpected insights, not easily obtainable by other methods. Mathematical concepts that have no direct connection with the real world often have deeper, indirect connections. It is those hidden links that make mathematics so useful.

CHAPTER 17

The Shape of Logic

Putting mathematics on fairly firm foundations

As the superstructure of mathematics grew ever larger, a small number of mathematicians began to wonder whether the foundations could support its weight. A series of foundational crises – in particular the controversies over the basic concepts of calculus and the general confusion about Fourier series – had made it clear that mathematical concepts must be defined very carefully and precisely to avoid logical pitfalls. Otherwise the subject's towers of deduction could easily collapse in logical contradictions, because of some underlying vagueness or ambiguity.

At first, such worries were focused on complicated, sophisticated ideas, such as Fourier series. But slowly the mathematical world came to realize that very basic ideas might also be suspect. Paramount among them was the concept of a number. The dreadful truth was that mathematicians had devoted so much effort to the discovery of deep properties of numbers that they had neglected to ask what numbers *were*. And when it came to giving a logical definition, they didn't know.

Dedekind

In 1858, teaching a calculus course, Dedekind became worried about the basis of calculus. Not its use of limits, but the system of real numbers. He published his thoughts in 1872 as *Stetigkeit und Irrationale Zahlen*, pointing out that apparently obvious properties of real numbers had never been proved in any rigorous way. As an example he cited the equation $\sqrt{2}\ \sqrt{3} = \sqrt{6}$. Obviously this fact follows by squaring both sides of the equation – except that multiplication of irrational numbers had never actually been defined. In his 1888 book *Was Sind und was Sollen die Zahlen?* (*What are Numbers, and What do They Mean?*) he exposed serious gaps in the logical foundations of the system of real numbers. No one had really proved that the real numbers exist.

He also proposed a way to fill these gaps, using what we now call *Dedekind cuts*. The idea was to start from an established number system, the rational numbers, and then to extend this system to obtain the richer system of real numbers. His approach was to start from the properties required of real numbers, find some way to rephrase them solely in terms of rational numbers and then reverse the procedure, interpreting those features of rational numbers as a definition of the reals. This kind of reverse engineering of new concepts from old ones has been widely used ever since.

Suppose, for the moment, that real numbers do exist. How do they relate to rational numbers? Some reals are not rational, an obvious example being $\sqrt{2}$. Now, although it is not an exact fraction, it can be approximated as closely as we wish by rationals. It somehow sits in a specific position, sandwiched among the dense array of all possible rationals. But how can we specify that position? Dedekind realized that $\sqrt{2}$ neatly separates the set of rational numbers into two pieces: those that are less than $\sqrt{2}$, and those that are greater. In a sense, this separation – or cut – defines the number $\sqrt{2}$ in terms of rationals. The only snag is that we make use of $\sqrt{2}$ in order to define the two pieces of the cut. However, there is a way out. The rational numbers greater than $\sqrt{2}$ are precisely those that are positive, and for which the square is greater than 2. The rational numbers less than $\sqrt{2}$ are all the others. These two sets of rational numbers are now defined without any *explicit* use of $\sqrt{2}$, but they specify its location on the real-number line precisely.

Dedekind showed that if, for the sake of argument, we assume that the real numbers exist, then a cut satisfying these two properties can be associated with any real number, by forming two sets: the set R of all rationals that are larger than that real number, and the set L of all rationals that are smaller than that real number, or equal to it. (The final condition is needed to associate a cut with any *rational* number. We don't want to leave them out.) Here L and R can be read as left and right in the usual picture of the real number line.

These two sets L and R obey some rather stringent conditions. First, every rational number belongs to precisely one of them. Second, every number in R is greater than any number in L. Finally, there is a technical condition that takes care of the rational numbers themselves: L may or may not have a largest member, but R never has a smallest member. Call *any* pair of subsets of the rationals with these properties a *cut*.

In reverse engineering, we do not need to assume that real numbers exist. Instead, we can use cuts to *define* real numbers, so that effectively a real number is a cut. We don't usually think of a real number that way, but Dedekind realized that we can do so if we wish. The main task is to define how to add and multiply cuts, so that the arithmetic of real numbers makes sense. This turns out to be easy. To add two cuts (L_1, R_1) and (L_2, R_2), define $L_1 + L_2$ to be the set of all numbers obtainable by adding a number in L_1 to a number in L_2 and similarly define $R_1 + R_2$. Then the sum of the two cuts is the cut $(L_1 + L_2, R_1 + R_2)$. Multiplication is similar, but positive and negative numbers behave slightly differently.

Finally, we have to verify that the arithmetic of cuts has all the features that we expect of real numbers. These include the standard laws of algebra, which all follow from analogous features of rational numbers. The crucial feature, which distinguishes the reals from the rationals, is that the limit of an infinite sequence of cuts exists (under certain technical conditions). Equivalently, there is a cut corresponding to any infinite decimal expansion. This is also fairly straightforward.

Assuming all this can be done, let's see how Dedekind can then prove that $\sqrt{2}\,\sqrt{3} = \sqrt{6}$. We have seen that $\sqrt{2}$ corresponds to the cut (L_1, R_1) where R_1 consists of all positive rationals the square of which is greater than 2. Similarly, $\sqrt{3}$ corresponds to the cut (L_2, R_2) where R_2 consists of all positive rationals the square of which is greater than 3. The product of these cuts is easily proved to be (L_3, R_3) where R_3 consists of all positive rationals the square of which is greater than 6. But this is the cut corresponding to $\sqrt{6}$. Done!

The beauty of Dedekind's approach is that it reduces all issues concerning real numbers to corresponding issues about rational numbers – specifically, about *pairs* of *sets* of rational numbers. It therefore defines real numbers purely in terms of rational numbers and operations on those numbers. The upshot is that real numbers exist (in a mathematical sense) provided rational numbers do.

There is a small price to pay: a real number is now defined as a pair of sets of rationals, which is not how we usually think of a real number. If this sounds strange, bear in mind that the usual representation of a real number as an infinite decimal requires an infinite sequence of decimal digits 0–9. This is conceptually at least as complicated as a Dedekind cut. But it is actually very tricky to define the sum or product of two infinite decimals, because the usual arithmetical methods for adding or multiplying decimals start from the right-hand end – and when a decimal is infinite, it doesn't *have* a right-hand end.

Axioms for whole numbers

Dedekind's book was all very well as a foundational exercise, but as the general point about defining your terms sank in, it was soon realized that the book had merely shifted attention from the reals to the rationals. How do we know that *rational* numbers exist? Well, if we assume the integers exist, this is easy: define a rational p/q to be a pair of integers (p, q) and work out the formulas for sums and products. If integers exist, so do pairs of integers.

Yes, but how do we know that integers exist? Apart from a plus or minus sign, integers are ordinary whole numbers. Taking care of the signs is easy. So, integers exist provided whole numbers do.

We still haven't finished, though. We are so familiar with whole numbers that it never occurs to us to ask whether the familiar numbers 0, 1, 2, 3 and so on actually exist. And if they do, *what are they*?

In 1889, Giuseppe Peano sidestepped the question of existence by taking a leaf out of Euclid's book. Instead of discussing the existence of points, lines, triangles and the like, Euclid simply wrote down a list of axioms – properties that would be assumed without further question. Never mind whether points and so on *exist* – a more interesting question is: if they did, what properties would follow? So Peano wrote down a list of axioms for whole numbers. The main features were:

- There exists a number 0.
- Every number n has a successor, $s(n)$ (which we think of as $n + 1$).
- If $P(n)$ is a property of numbers, such that $P(0)$ is true, and whenever $P(n)$ is true then $P(s(n))$ is true, then $P(n)$ is true for every n (Principle of Mathematical Induction).

He then defined the numbers 1, 2 and so on in terms of those axioms, essentially by setting

$$1 = s(0)$$

$$2 = s(s(0))$$

and so on. And he also defined the basic operations of arithmetic and proved that they obey the usual laws. In his system, $2 + 2 = 4$ is a provable theorem, stated as $s(s(0)) + s(s(0)) = s(s(s(s(0))))$.

A great advantage of this axiomatic approach is that it pins down exactly what we have to prove if we want to show, by some means or other, that whole numbers exist. We just have to construct some system that satisfies all of Peano's axioms.

The deep question here is the meaning of 'exist' in mathematics. In the real world, something exists if you can observe it, or, failing that, infer its necessary presence from things that can be observed. We know that gravity exists because we can observe its effects, even though no one can *see* gravity. So, in the real world, we can sensibly talk about the existence of two cats, two bicycles or two loaves of bread. However, the *number* two is not like that. It is not a thing, but a conceptual construct. We never meet the number two in the real world. The closest we get is a symbol, 2, written or printed on paper, or displayed on a computer screen. However, no one imagines that a symbol is the same as the thing it represents. The word 'cat' written in ink is not a cat. Similarly, the symbol 2 is not the number two.

The meaning of 'number' is a surprisingly difficult conceptual and philosophical problem. It is made all the more frustrating by the fact that we all know perfectly well how to use numbers. We know how they behave, but not what they are.

Sets and classes

In the 1880s Gottlob Frege tried to resolve this conceptual issue by constructing whole numbers out of even simpler objects – namely sets, or classes as he called them. His starting point was the standard association of numbers with counting. According to Frege, two is a property of those sets – and only those – that can be matched one-to-one with a standard set $\{a, b\}$ having distinct members a and b. So

{one cat, another cat}

{one bicycle, another bicycle}

{one loaf, another loaf}

can all be matched with $\{a, b\}$, so they all determine – whatever that means – the same *number*.

Unfortunately, using a list of standard sets as numbers seems to beg the question – it's very like confusing a symbol with what it represents. But how can we characterize 'a property of those sets that can be matched one-to-one with a standard set'? What is a property? Frege had a wonderful insight. There is a well-defined set that is associated with any property; namely, the set consisting of everything that possesses that property. The property 'prime' is associated with the set of *all* prime numbers; the property 'isosceles' is associated with the set of *all* isosceles triangles, and so on.

So Frege proposed that number two is the set comprising *all* sets that can be matched one-to-one with the standard set $\{a, b\}$. More generally, a *number* is the set of all sets that can be matched to any given set. So, for example, the number 3 is the set

{... $\{a, b, c\}$, {one cat, another cat, yet another cat}, {X, Y, Z}, ...}

although it is probably best to use mathematical objects in place of cats or letters.

On this basis, Frege discovered that he could put all of the arithmetic of the whole numbers on a logical basis. It all reduced to obvious properties of sets. He wrote everything up in his masterwork *Die Grundlagen der Arithmetik* (*The Foundations of Arithmetic*) of 1884, but to his bitter disappointment Georg Cantor, a leading mathematical logician, dismissed the book as worthless. In 1893

Frege, undaunted, published the first volume of another book, *Die Grundgesetze der Arithmetik* (*The Basic Laws of Arithmetic*), in which he provided an intuitively plausible system of axioms for arithmetic. Peano reviewed it, and everyone else ignored it. Ten years later, Frege was finally ready to publish volume two, but by then he had noticed a basic flaw in his axioms. Others noticed this as well. While volume two was in press, disaster struck. Frege received a letter from the philosopher-mathematician Bertrand Russell, to whom he had sent an advance copy of his book. To paraphrase, the letter said roughly this: 'Dear Gottlob, consider the set of all sets that are not members of themselves. Yours, Bertrand.'

Frege was a superb logician and he immediately got Russell's point – indeed, he was already aware of the potential for trouble. Frege's whole approach had assumed, without proof, that any

Russell's Paradox

A less formal version of the paradox proposed by Russell is the village barber, who shaves everyone who does not shave themselves. Who shaves the barber? If he does shave himself, then by definition he is shaved by the village barber – himself! If he does not shave himself, then he is shaved by the barber – which, again, is himself.

Aside from various cooks – the barber is a woman, for instance – the only possible conclusion is that no such barber exists. Russell reformulated this paradox in terms of sets. Define a set *X* to consist of all sets that are not members of themselves. Is *X* a member of itself, or not? If it is not, then by definition it belongs to *X* – itself. If it is a member of itself, then like all members of *X*, it is not a member of itself. So, X is a member of itself if it is not, and it is not a member of itself it is. This time there is no way out – female sets are not yet part of the mathematical enterprise.

reasonable property defined a meaningful set, consisting of those objects that possess the property concerned. But here was an apparently reasonable property, not a member of itself, which manifestly did *not* correspond to a set.

A glum Frege penned an appendix to his magnum opus, discussing Russell's objection. He found a short-term fix: eliminate from the realm of sets any that are members of themselves. But he was never really happy with this proposal.

Russell, for his part, tried to repair the gap in Frege's construction of whole numbers from sets. His idea was to restrict the kind of property that could be used to define a set. Of course he had to find a proof that this restricted kind of property never led to a paradox. In collaboration with Alfred North Whitehead, he came up with a complicated and technical *theory of types* which achieved that objective, to their satisfaction at least. They wrote their approach in a massive three-volume tome *Principia Mathematica* of 1910–13. The definition of the number 2 is near the end of volume one, and the theorem $1 + 1 = 2$ is proved on page 86 of volume two. However, *Principia Mathematica* did not end the foundational debate. The theory of types was itself contentious. Mathematicians wanted something simpler and more intuitive.

Cantor

These analyses of the fundamental role of counting as the basis for numbers led to one of the most audacious discoveries in the whole of mathematics: Cantor's theory of *transfinite numbers* – different sizes of infinity.

Infinity, in various guises, seems unavoidable in mathematics. There is no largest whole number – because adding one always produces a larger number still – so there are infinitely many whole numbers. Euclid's geometry takes place on an infinite plane, and he proved that there are infinitely many prime numbers. In the run-up to calculus, several people, among them Archimedes, found it useful to think of an area or a volume as being the sum of infinitely many infinitely thin slices. In the aftermath of calculus, the same picture of areas and volumes was used for heuristic purposes, even if the actual proofs took a different form.

These occurrences of the infinite could be rephrased in finite terms to avoid various philosophical difficulties. Instead of saying 'there are infinitely many whole numbers', for instance, we can instead say 'there is no largest whole number'. The second statement avoids explicit mention of the infinite, while being logically equivalent to the first. Essentially, infinity is here being thought of as a process, which can be continued without any specific limit, but is not actually *completed*. Philosophers call this kind of infinity potential infinity. In contrast, explicit use of infinity as a mathematical object in its own right is actual infinity.

Mathematicians prior to Cantor had noticed that actual infinities had paradoxical features. In 1632 Galileo wrote his *Dialogue Concerning the Two Chief World Systems*, in which two fictional characters, the sagacious Salviati and the intelligent layman Sagredo, discuss the causes of tides, from the geocentric and heliocentric viewpoints. All mention of tides was removed at the request of church authorities, making the book a hypothetical exercise which nonetheless makes a powerful case for Copernicus's heliocentric theory. Along the way, the two characters discuss some of the paradoxes of infinity. Sagredo asks 'are there more numbers than squares?' and points out that since most whole numbers are not perfect squares, the answer must be yes. Salviati replies that every number can be matched uniquely to its square:

$$\begin{array}{cccccccc} 1 & 2 & 3 & 4 & 5 & 6 & 7 & \dots \\ \downarrow & \downarrow & \downarrow & \downarrow & \downarrow & \downarrow & \downarrow & \\ 1 & 4 & 9 & 16 & 25 & 36 & 49 & \dots \end{array}$$

Therefore the number of whole numbers must be the same as that of squares, so the answer is no.

Cantor resolved this difficulty by recognizing that in the dialogue, the adjective 'more' is being used in two different ways. Sagredo is pointing out that the set of all squares is a proper subset of the set of all whole numbers. Salviati's position is subtler: he is arguing that there is a one-to-one correspondence between the set of all squares and the set of all whole numbers. Those statements are different, and both can be true without leading to any contradiction.

Pursuing this line of thought, Cantor was led to the invention of an arithmetic of the infinite, which explained the previous paradoxes while introducing some new ones. This work was part of a more extensive programme, *Mengenlehre*, the mathematics of sets (Menge in German means set or assemblage). Cantor began thinking about sets because of some difficult questions in Fourier analysis, so the ideas were rooted in conventional mathematical theories. But the answers he discovered were so strange that many mathematicians of the period rejected them out of hand. Others, however, realized their value, notably David Hilbert, who affirmed 'No one shall expel us from the paradise that Cantor has created'.

Set size

Cantor's starting point was the naive concept of a *set*, which is a collection of objects, its *members*. One way to specify a set is to list the members, using curly brackets. For instance, the set of all whole numbers between 1 and 6 is written

$$\{1, 2, 3, 4, 5, 6\}$$

Alternatively, a set can be specified by stating the rule for membership:

$$\{n : 1 \leq n \leq 6 \text{ and } n \text{ is a whole number}\}$$

The sets specified above are identical. The first notation is limited to finite sets, but the second has no such limitations. Thus the sets

$$\{n : n \text{ is a whole number}\}$$

and

$$\{n : n \text{ is a perfect square}\}$$

are both specified precisely, and both are infinite.

One of the simplest things you can do with a set is count its members. How big is it? The set {1, 2, 3, 4, 5, 6} has six members. So does the set {1, 4, 9, 16, 25, 36} consisting of the corresponding squares. We say that the *cardinality* of the set is 6, and call 6 a *cardinal number*. (There is a different concept, ordinal number, associated with putting numbers in order, which is why the adjective 'cardinal' is not superfluous here.) The set of all whole numbers cannot be counted in this manner, but Cantor noticed that despite that, you can place the set of all whole numbers and the set of all squares in one-to-one correspondence, using the same scheme as Galileo. Each whole number n is paired with its square n^2.

Cantor defined two sets to be *equinumerous* (not his word) if there exists a one-to-one correspondence between them. If the sets are finite, this property is equivalent to 'having the same number of members'. But if the sets are infinite, it apparently makes no sense to talk of the number of members; nevertheless the concept of equinumerosity makes perfect sense. But Cantor went further. He introduced a system of *transfinite numbers*, or *infinite cardinals*, which did make it possible to say how many members an infinite set has. Moreover, two sets were equinumerous if and only if they had the same number of members – the same cardinal.

The starting point was a new kind of number, which he denoted by the symbol $\aleph_0$. This is the Hebrew letter aleph with a subscript zero, read as aleph-null in German, nowadays aleph-zero. This number is defined to be the cardinality of the set of all whole numbers. By insisting that equinumerous sets have the same cardinality, Cantor then required any set that can be placed in one-to-one correspondence with the set of all whole numbers also to have cardinality $\aleph_0$. For instance, the set of all squares has cardinality $\aleph_0$. So does the set of all even numbers:

$$\begin{array}{cccccccc} 1 & 2 & 3 & 4 & 5 & 6 & 7 & \dots \\ \downarrow & \downarrow & \downarrow & \downarrow & \downarrow & \downarrow & \downarrow & \\ 2 & 4 & 6 & 8 & 10 & 12 & 14 & \dots \end{array}$$

And so does the set of all odd numbers:

$$\begin{array}{cccccccc} 1 & 2 & 3 & 4 & 5 & 6 & 7 & \dots \\ \downarrow & \downarrow & \downarrow & \downarrow & \downarrow & \downarrow & \downarrow & \\ 1 & 3 & 5 & 7 & 9 & 11 & 13 & \dots \end{array}$$

One implication of these definitions is that a smaller set can have the same cardinality as a bigger one. But there is no logical contradiction here to Cantor's definitions, so he considered this feature to be a natural consequence of his set-up, and a price worth paying. You just have to be careful not to assume that infinite cardinals behave just like finite ones. But why should they? They are not finite!

Are there more integers (positive and negative) than whole numbers? Aren't there twice as many? No, because we can match the two sets like this:

$$\begin{array}{cccccccc} 1 & 2 & 3 & 4 & 5 & 6 & 7 & \dots \\ \downarrow & \downarrow & \downarrow & \downarrow & \downarrow & \downarrow & \downarrow & \\ 0 & 1 & -1 & 2 & -2 & 3 & -3 & \dots \end{array}$$

The arithmetic of infinite cardinals is also strange. For instance, we have just seen that the sets of even and odd whole numbers have cardinal $\aleph_0$. Since these sets have no members in common, the cardinal of their union – the set formed by combining them – should, by analogy with finite sets, be $\aleph_0 + \aleph_0$. But we know what that union is: it is the whole numbers, with cardinal $\aleph_0$. So apparently we are forced to deduce that

$$\aleph_0 + \aleph_0 = \aleph_0.$$

And so we are. But again, there is no contradiction: we cannot divide by $\aleph_0$, to deduce that $1 + 1 = 1$, because $\aleph_0$ is not a whole number and division has not been defined, let alone been shown to make sense. Indeed, this equation shows that division by $\aleph_0$ does not always make sense. Again, we accept this as the price of progress.

This is all very well, but it looks as though $\aleph_0$ is just a fancy symbol for good old ∞, and nothing new is really being said. Is it not the case that *all* infinite sets have cardinal $\aleph_0$? Surely all infinities are equal?

One candidate for an infinite cardinal greater than $\aleph_0$ – that is, an infinite set that cannot be put in one-to-one correspondence with the set of all whole numbers – is the set of all rational numbers, which is usually denoted by $\mathbb{Q}$. After all, there are infinitely many rational numbers in the gap between any two consecutive integers, and the kind of trick we used for the integers no longer works.

However, in 1873 Cantor proved that $\mathbb{Q}$ also has cardinal $\aleph_0$. The one-to-one correspondence shuffles the numbers quite thoroughly, but no one said that they had to remain in numerical order. It was beginning to look remarkably as though every infinite set had cardinal $\aleph_0$.

In the same year, though, Cantor made a breakthrough. He proved that the set $\mathbb{R}$ of all real numbers does *not* have cardinal $\aleph_0$, an astonishing theorem that he published in 1874. So even in Cantor's special sense, there are *more* reals than integers. One infinity can be bigger than another infinity.

How big is the cardinal of the reals? Cantor hoped that it would be $\aleph_1$, the next largest cardinal after $\aleph_0$. But he could not prove this, so he named the new cardinal c, for continuum. The hoped-for equation $c = \aleph_1$ was called the continuum hypothesis. Only in 1960 did mathematicians sort out the relationship between c and $\aleph_1$, when Paul Cohen proved that the answer depends on which axioms you choose for set theory. With some sensible axioms, the two cardinals are the same. But with other, equally reasonable axioms, they are different.

Although the validity of the equation $c = \aleph_1$ depends on the chosen axioms, an associated equality does not. This is $c = 2^{\aleph_0}$. For any cardinal A we can define 2^A as the cardinal of the set of all subsets of A. And we can prove, very easily, that 2^A is always bigger than A. This means that not only are some infinities bigger than others: there is no largest infinite cardinal.

Contradictions

The biggest task of foundational mathematics, however, was not proving that mathematical concepts exist. It was to prove that mathematics is logically consistent. For all mathematicians knew – indeed, for all they know today – there might be some sequence of logical steps, all perfectly correct, leading to an absurd conclusion. Maybe you could prove that $2 + 2 = 5$, or $1 = 0$, for instance. Or that 6 is prime, or $\pi = 3$.

Now, it might seem that one tiny contradiction would have limited consequences. In everyday life, people often operate very happily within a contradictory framework, asserting at one moment that, say, global warming is destroying the planet, and a moment later that low-cost airlines are a great invention. But in mathematics, consequences are not limited, and you can't evade logical contradictions by ignoring them. In mathematics, once something is proved you can use it in other proofs. Having proved $0 = 1$, much nastier things follow. For instance, all numbers are equal. For if x is any number, start from $0 = 1$ and multiply both sides by x. Then $0 = x$. Similarly, if y is any other number, $0 = y$. But now $x = y$.

Worse, the standard method of proof by contradiction means that anything can be proved once we have proved that $0 = 1$. To prove Fermat's Last Theorem, say, we argue like this:

Suppose Fermat's Last Theorem is false.
Then (as already proved) $0 = 1$.
Contradiction.
Therefore Fermat's Last Theorem is true.

As well as being unsatisfying, this method also proves Fermat's Last Theorem is false:

Suppose Fermat's Last Theorem is true.
Then (as already proved) $0 = 1$.
Contradiction.
Therefore Fermat's Last Theorem is false.

Once everything is true – and also false – nothing meaningful can be said. The whole of mathematics would be a silly game, with no content.

Hilbert

The next major foundations step was taken by David Hilbert, probably the leading mathematician of his day. Hilbert had a habit of working in one area of mathematics for about ten years, polishing off the main problems, and moving on to a new area. Hilbert became convinced that it should be possible to prove that mathematics can never lead to a logical contradiction. He also realized that physical intuition would not be useful in such a project. If mathematics is contradictory, it must be possible to prove that $0 = 1$, in which case there is a physical interpretation: 0 cows = 1 cow, so cows can disappear in a puff of smoke. This seems unlikely. However, there is no guarantee that the mathematics of whole numbers actually matches the physics of cows, and it is at least conceivable that a cow might suddenly disappear. (In quantum mechanics, this could happen, but with a very low probability.) There is a limit to the number of cows in a finite universe, but there is no limit to the size of mathematical integers. So physical intuition could be misleading, and should be ignored.

David Hilbert

1862–1943

David Hilbert graduated from the University of Königsberg in 1885 with a thesis on invariant theory. He was on the university's staff until taking up a professorship at Göttingen in 1895. He continued to work in invariant theory, proving his finite basis theorem in 1888. His methods were more abstract than the prevailing fashion, and one of the leading figures in the field, Paul Gordan, found the work unsatisfactory. Hilbert revised his paper for publication in the *Annalen*, and Klein called it 'the most important work on general algebra that [the journal] has ever published.'

In 1893 Hilbert began a comprehensive report on number theory, the *Zahlbericht*. Although this was intended to summarize the known state of the theory, Hilbert included a mass of original material, the basis of what we now call class field theory.

By 1899 he had changed fields again, now studying the axiomatic foundations of Euclidean geometry. In 1900, at the Second International Congress of Mathematicians in Paris, he presented a list of 23 major unsolved problems. These *Hilbert problems* had a tremendous effect on the subsequent direction of mathematical research.

Around 1909 his work on integral equations led to the formulation of *Hilbert spaces*, now basic to quantum mechanics. He also came very close to discovering Einstein's equations for general relativity in a 1915 paper. He added a note in proof to the effect that the paper was consistent with Einstein's equations, which gave rise to an erroneous belief that Hilbert might have anticipated Einstein.

In 1930, on his retirement, Hilbert was made an honorary citizen of Königsberg. His acceptance speech ended with the words 'Wir müssen wissen, wir werden wissen' (we must know, we shall know) which encapsulated his belief in the power of mathematics and his determination to solve even the most difficult problems.

Hilbert came to this point of view in his work on the axiomatic basis of Euclid's geometry. He discovered logical flaws in Euclid's axiom system, and realized that these flaws had arisen because Euclid had been misled by his visual imagery. Because he knew that a line was a long thin object, a circle was round and a point was a dot, he had inadvertently assumed certain properties of these objects, without stating them as axioms. After several attempts, Hilbert put forward a list of 21 axioms and discussed their role in Euclidean geometry in his 1899 *Grundlagen der Geometrie* (*Foundations of Geometry*).

Hilbert maintained that a logical deduction must be valid, independently of the interpretation imposed on it. Anything that relies on some particular interpretation of the axioms, but fails in other interpretations, involves a logical error. It is this view of axiomatics, rather than the specific application to geometry, that is Hilbert's most important influence on the foundations of mathematics. In fact, the same point of view also influenced the content of mathematics, by making it much easier – and more respectable – to invent new concepts by listing axioms for them. Much of the abstraction of early 20th century mathematics stemmed from Hilbert's viewpoint.

It is often said that Hilbert advocated the idea that mathematics is a meaningless game played with symbols, but this overstates his position. His point was that in order to place the subject on a firm logical basis, you have to think about it *as* if it is a meaningless game played with symbols. All else is irrelevant to the logical structure. But no one who takes a serious look at Hilbert's mathematical discoveries, and his deep commitment to the subject, can reasonably deduce that he thought he was playing a meaningless game.

After his success in geometry, Hilbert now set his sights on a far more ambitious project: to place the whole of mathematics on a sound logical footing. He followed the work of the leading logicians closely, and developed an explicit programme to sort out the foundations of mathematics once and for all. As well as proving that mathematics was free of contradictions, he also believed that in principle every problem could be solved – every mathematical statement could either be proved or disproved. A number of early successes convinced him that he was heading along the correct path, and that success was not far away.

Gödel

There was one logician, however, who remained unconvinced by Hilbert's proposal to prove that mathematics is logically consistent. His name was Kurt Gödel, and his worries about Hilbert's programme forever changed our view of mathematical truth.

Before Gödel, mathematics was simply though to be true – and it was the highest example of truth, because the truth of a statement like 2 + 2 = 4 was something in the realm of pure thought, independent of our physical world. Mathematical truths were not

What logic did for them

Charles Lutwidge Dodgson, better known as Lewis Carroll, used his own formulation of a branch of mathematical logic, now known as propositional calculus, to set and solve logical puzzles. A typical example from his *Symbolic Logic* of 1896 is:

- Nobody, who really appreciates Beethoven, fails to keep silence while the Moonlight Sonata is being played;
- Guinea-pigs are hopelessly ignorant of music;
- No one, who is hopelessly ignorant of music, ever keeps silence while the Moonlight Sonata is being played.

The deduction is that no guinea-pig really appreciates Beethoven. This form of logical argument is called a syllogism, and it goes back to classical Greece.

things that might be disproved by later experiments. In this they were superior to physical truths, such as Newton's inverse square law of gravity, which was disproved by observations of the motion of the perihelion of Mercury, supporting the new gravitational theory suggested by Einstein.

After Gödel, mathematical truth turned out to be an illusion. What existed were mathematical *proofs*, the internal logic of which might well be faultless, but which existed in a wider context – foundational mathematics – where there could be no guarantee that the entire game had any meaning at all. Gödel did not just assert this: he proved it. In fact, he did two things, which together left Hilbert's careful, optimistic programme in ruins.

Gödel proved that if mathematics is logically consistent, then it is impossible to prove that. Not just that he could not find a proof, but that *no proof exists*. So, remarkably, if you do succeed in proving that mathematics is consistent, it immediately follows that it's not. He also proved that some mathematical statements can neither be proved nor disproved. Again, not just that he personally could not achieve this, but that it is *impossible*. Statements of this type are called *undecidable*.

He proved these statements initially within a particular logical formulation of mathematics, that adopted by Russell and Whitehead in their *Principia Mathematica*. To begin with, Hilbert thought that there might be a way out: find a better foundation. But as the logicians studied Gödel's work, it quickly became apparent that the same ideas would work in *any* logical formulation of mathematics strong enough to express the basic concepts of arithmetic.

An intriguing consequence of Gödel's discoveries is that any axiomatic system for mathematics must be *incomplete*: you can never write down a finite list of axioms that will determine all true and false theorems uniquely. There was no escape: Hilbert's programme cannot work. It is said that when Hilbert first heard of Gödel's work he was extremely angry. His anger may well have been

Kurt Gödel

1906–78

In 1923, when Gödel went to the University of Vienna, he was still unsure whether to study mathematics or physics. His decision was influenced by lectures of a severely disabled mathematician, Philipp Furtwängler (brother of the famous conductor and composer Wilhelm). Gödel's own health was fragile, and Furtwängler's will to overcome his disabilities made a big impression. In a seminar given by Moritz Schlick, Gödel began studying Russell's *Introduction to Mathematical Philosophy*, and it became clear that his future lay in mathematical logic.

His doctoral thesis of 1929 proved that one restricted logical system, first-order propositional calculus, is complete – every true theorem can be proved and every false one can be disproved. He is best known for his proof of 'Gödel's Incompleteness Theorems'. In 1931 Gödel published his epic paper *Über formal unentscheidbare Sätze der Principia Mathematica und verwandter Systeme*. In it, he proved that no system of axioms rich enough to formalize mathematics can be logically complete. In 1931 he discussed his work with the logician Ernst Zermelo, but the meeting fared badly, possibly because Zermelo had already made similar discoveries but had failed to publish them.

In 1936 Schlick was murdered by a Nazi student, and Gödel had a mental breakdown (his second). When he had recovered, he visited Princeton. In 1938 he married Adele Porkert, against his mother's wishes, and returned to Princeton, shortly after Austria had been incorporated into Germany. When the Second World War started, he was worried that he might be called up into the German Army, so he emigrated to the USA, travelling through Russia and Japan. In 1940 he produced a second seminal work, a proof that Cantor's continuum hypothesis is consistent with the usual axioms for mathematics.

He became a US citizen in 1948, and spent the rest of his life in Princeton. Towards the end of his life he worried more and more about his health, and eventually became convinced that someone was trying to poison him. He refused food, and died in hospital. To the end, he liked to argue philosophy with his visitors.

directed at himself, because the basic idea in Gödel's work is straightforward. (The technical implementation of that idea is distinctly difficult, but Hilbert was good with technicalities.) Hilbert probably realized that he should have seen Gödel's theorems coming.

Russell demolished Frege's book with a logical paradox, the paradox of the village barber who shaves everyone who does not shave himself: the set of all sets that are not members of themselves. Gödel demolished Hilbert's programme with another logical paradox, the paradox of someone who says: this statement is a lie. For in effect Gödel's undecidable statement – on which all else rests – is a theorem T that states: 'this theorem cannot be proved'.

If every theorem can either be proved, or disproved, then Gödel's statement T is contradictory in both cases. Suppose T can be proved: then T states that T cannot be proved, a contradiction. On the other hand, if T can be disproved, then the statement T is false, so it is wrong to state that T cannot be proved. Therefore T can be proved, another contradiction. So the assumption that every theorem can either be proved or disproved tells us that T can be proved if and only if it cannot be proved.

Where are we now?

Gödel's theorems changed the way we view the logical foundations of mathematics. They imply that currently unsolved problems may have no solution at all –neither true nor false, but in the limbo of undecidability. And many interesting problems have been shown to be undecidable. However, the effect of Gödel's work has not, in practice, extended much beyond the foundational area where it took place. Rightly or wrongly, mathematicians working on the Poincaré conjecture, or the Riemann hypothesis, spend their time either looking for proofs or disproofs. They are aware that the problem may be undecidable, and might even look for a proof of undecidability if they could see how to get started. But most known undecidable problems have a self-referential feel to them, and without that, a proof of undecidability seems unattainable anyway.

As mathematics built ever more complicated theories on top of earlier ones, the superstructure of mathematics began to come to pieces because of unrecognized assumptions that turned out to be false. In order to shore up the entire edifice, serious work was needed on the foundations.

The subsequent investigations delved into the true nature of numbers, working backwards from complex numbers to reals, to rationals and then to whole numbers. But the process did not stop there. Instead, the number systems themselves were reinterpreted in terms of even simpler ingredients, sets.

Set theory led to major advances, including a sensible, though unorthodox, system of infinite numbers. It also revealed some fundamental paradoxes related to the notion of a set. The resolution of these paradoxes was not, as Hilbert hoped, a complete vindication of axiomatic mathematics, and a proof of its logical consistency. Instead, it was a proof that mathematics has inherent limitations, and that some problems *do not have solutions*. The upshot was a profound change in the way we think about mathematical truth and certainty. It is better to be aware of our limitations than to live in a fool's paradise.

What logic does for us

A profound variant on Gödel's incompleteness theorem was discovered by Alan Turing in an analysis of which computations are feasible, published in 1936 as *On Computable Numbers, with an application to the Entscheidungsproblem (The decision problem)*. Turing began by formalizing an algorithmic computation – one that follows a pre-stated recipe – in terms of a so-called Turing machine. This is a mathematical idealization of a device which writes symbols 0 and 1 on a moving tape according to specific rules. He proved that the halting problem for Turing machines – does the computation eventually stop for a given input? – is *undecidable*. This means that there is no algorithm that can predict whether the computation halts or not.

Turing proved his result by assuming the halting problem to be decidable and constructing a computation that halts if and only if it does not halt, a contradiction. His result demonstrates that there are limits to computability. Some philosophers have extended these ideas to determine limits on rational thinking, and it has been suggested that a conscious mind cannot function algorithmically. However, the arguments here are currently inconclusive. They do show that it is naïve to think that a brain works much like a modern computer, but this may not imply that a computer cannot simulate a brain.

CHAPTER 18

How Likely is That?

The rational approach to chance

The growth of mathematics in the 20th and early 21st centuries has been explosive. More new mathematics has been discovered in the last 100 years than in the whole of previous human history. Even to sketch these discoveries would take thousands of pages, so we are forced to look at a few samples from the enormous amount of material that is available.

An especially novel branch of mathematics is the theory of probability, which studies the chances associated with random events. It is the mathematics of uncertainty. Earlier ages scratched the surface, with combinatorial calculations of odds in gambling games and methods for improving the accuracy of astronomical observations despite observational errors, but only in the early 20th century did probability theory emerge as a subject in its own right.

Probability and statistics

Nowadays, probability theory is a major area of mathematics, and its applied wing, statistics, has a significant effect on our everyday lives – possibly more significant than any other single major branch of mathematics. Statistics is one of the main analytic techniques of the medical profession. No drug comes to market, and no treatment is permitted in a hospital, until clinical trials have ensured that it is sufficiently safe, and that it is effective. Safety here is a relative concept: treatments may be used on patients suffering from an otherwise fatal disease when the chance of success would be far too small to use them in less serious cases.

Probability theory may also be the most widely misunderstood, and abused, area of mathematics. But used properly and intelligently, it is a major contributor to human welfare.

Games of chance

A few probabilistic questions go back to antiquity. In the Middle Ages, we find discussions of the chances of throwing various numbers with two dice. To see how this works, let's start with one die. Assuming that the die is fair – which turns out to be a difficult concept to pin down – each of the six numbers 1, 2, 3, 4, 5 and 6 should be thrown equally often, in the long run. In the short run, equality is impossible: the first throw must result in just one of those numbers, for instance. In fact, after six throws it is actually rather unlikely to have thrown each number exactly once. But in a long series of throws, or trials, we expect each number to turn up roughly one time in six; that is, with probability $1/6$. If this didn't happen, the die would in all likelihood be unfair or biased.

An event of probability 1 is certain, and one of probability 0 is impossible. All probabilities lie between 0 and 1, and the probability of an event represents the proportion of trials in which the event concerned happens.

Back to that medieval question. Suppose we throw two dice simultaneously (as in numerous games from craps to Monopoly™). What is the probability that their sum is 5? The upshot of numerous arguments, and some experiments, is that the answer is $1/9$. Why? Suppose we distinguish the two dice, colouring one blue and the other red. Each die can independently yield six distinct numbers, making a total of 36 possible pairs of numbers, all equally likely. The combinations (blue + red) that yield 5 are 1 + 4, 2 + 3, 3 + 2, 4 + 1; these are distinct cases because the blue die produces distinct results in each case, and so does the red one. So in the long run we expect to find a sum of 5 on four occasions out of 36, a probability of $\frac{4}{36} = \frac{1}{9}$.

Another ancient problem, with a clear practical application, is how to divide the stakes in a game of chance if the game is interrupted for some reason. The Renaissance algebraists Pacioli, Cardano, and Tartaglia all wrote on this question. Later, the Chevalier de Meré asked Pascal the same question, and Pascal and Fermat wrote each other several letters on the topic.

From this early work emerged an implicit understanding of what probabilities are and how to calculate them. But it was all very ill-defined and fuzzy.

Combinations

A working definition of the probability of some event is the proportion of occasions on which it will happen. If we roll a die, and the six faces are equally likely, then the probability of any particular face showing is $\frac{1}{6}$. Much early work on probability relied on calculating how many ways some event would occur, and dividing that by the total number of possibilities.

A basic problem here is that of combinations. Given, say, a pack of six cards, how many different subsets of four cards are there? One method is to list these subsets: if the cards are 1–6, then they are

1234 1235 1236 1245 1246

1256 1345 1346 1356 1456

2345 2346 2356 2456 3456

so there are 15 of them. But this method is too cumbersome for larger numbers of cards, and something more systematic is needed.

Imagine choosing the four members of the subset, one at a time. We can choose the first in six ways, the second in only five (since one has been eliminated), the third in four ways, the fourth in three ways. The total number of choices, in this order, is $6 \times 5 \times 4 \times 3 = 360$. However, every subset is counted 24 times – as well as 1234 we find 1243, 2134, and so on, and there are 24 ($4 \times 3 \times 2$) ways to rearrange four objects. So the correct answer is 360/24, which equals 15. This argument shows that the number of ways to choose m objects from a total of n objects is

$$\binom{n}{m} = \frac{n(n-1)\ldots(n-m+1)}{1 \times 2 \times 3 \times \cdots \times m}$$

These expressions are called *binomial coefficients*, because they also arise in the algebraic expansion of the binomial $(a+b)^n$. If we arrange them in a table, so that the nth row contains the binomial coefficients then the nth row looks like this:

$$\binom{n}{0}\binom{n}{1}\binom{n}{2}\cdots\binom{n}{n}$$

In the sixth row we see the numbers 1, 6, 15, 20, 15, 6, 1. Compare with the formula

$$(x+1)^6 = x^6 + 6x^5 + 15x^4 + 20x^3 + 15x^2 + 6x + 1$$

and we see the same numbers arising as the coefficients. This is not a coincidence.

The triangle of numbers (see next page) is called *Pascal's triangle* because it was discussed by Pascal in 1655. However, it was known much earlier; it goes back to about 950 in a commentary on an ancient Indian book called the *Chandas Shastra*. It was also known to the Persian mathematicians Al-Karaji and Omar Khayyám, and is known as the Khayyám triangle in modern Iran.

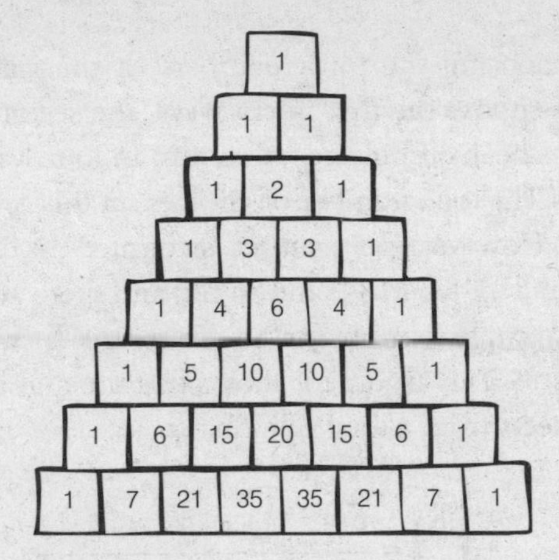

Pascal's triangle

Probability theory

Binomial coefficients were used to good effect in the first book on probability: the *Ars Conjectandi* (*Art of Conjecturing*) written by Jacob Bernoulli in 1713. The curious title is explained in the book:

> 'We define the art of conjecture, or stochastic art, as the art of evaluating as exactly as possible the probabilities of things, so that in our judgments and actions we can always base ourselves on what has been found to be the best, the most appropriate, the most certain, the best advised; this is the only object of the wisdom of the philosopher and the prudence of the statesman.'

So a more informative translation might be *The Art of Guesswork*.

Bernoulli took it for granted that more and more trials lead to better and better estimates of probability.

> 'Suppose that without your knowledge there are concealed in an urn 3000 white pebbles and 2000 black pebbles, and in trying to determine the numbers of these pebbles you take out one pebble after another (each time replacing the pebble...) and that you observe how often a white and how

> often a black pebble is drawn... Can you do this so often that it becomes ten times, one hundred times, one thousand times, etc, more probable... that the numbers of whites and blacks chosen are in the same 3:2 ratio as the pebbles in the urn, rather than a different ratio?'

Here Bernoulli asked a fundamental question, and also invented a standard illustrative example, balls in urns. He clearly believed that a ratio of 3:2 was the sensible outcome, but he also recognized that actual experiments would only approximate this ratio. But he believed that with enough trials, this approximation should become better and better.

There is a difficulty here, and it stymied the whole subject for a long time. In such an experiment, it is certainly *possible* that by pure chance every pebble removed could be white. So there is no hard and fast guarantee that the ratio must always tend to 3:2. The best we can say is that with very high probability, the numbers should approach that ratio. But now there is a danger of circular logic: we use ratios observed in trials to infer probabilities, but we also use probabilities to carry out that inference. How can we observe that the probability of all pebbles being white is very small? If we do that by lots of trials, we have to face the possibility that the result there is misleading, for the same reason; and the only way out seems to be even more trials to show that this event is itself highly unlikely. We are caught in what looks horribly like an infinite regress.

Fortunately, the early investigators of probability theory did not allow this logical difficulty to stop them. As with calculus, they knew what they wanted to do, and how to do it. Philosophical justification was less interesting than working out the answers.

Bernoulli's book contained a wealth of important ideas and results. One, the *Law of Large Numbers*, sorts out in exactly what sense long-term ratios of observations in trials correspond to probabilities. Basically, he proves that the probability that the ratio does *not* become very close to the correct probability tends to zero as the number of trials increases without limit.

Another basic theorem can be viewed in terms of repeatedly tossing a biased coin, with a probability p of giving heads, and $q = 1 - p$ of giving tails. If the coin is tossed twice, what is the probability of exactly 2, 1 or 0 heads? Bernoulli's answer was p^2, $2pq$ and q^2. These are the terms arising from the expansion of $(p + q)^2$ as $p^2 + 2pq + q^2$. Similarly if the coin is tossed three times, the probabilities of 3, 2, 1 or 0 heads are the successive terms in $(p + q)^3 = p^3 + 3p^2q + 3q^2p + q^3$.

More generally, if the coin is tossed n times, the probability of exactly m heads is equal to

$$\binom{m}{n} p^m q^{n-m}$$

the corresponding term in the expansion of $(p + q)^n$.

Between 1730 and 1738 Abraham De Moivre extended Bernoulli's work on biased coins. When m and n are large, it is difficult to work out the binomial coefficient exactly, and De Moivre derived an approximate formula, relating Bernoulli's binomial distribution to what we now call the *error function* or *normal distribution*

$$\frac{1}{\sqrt{2\pi}} e^{-x^2}$$

De Moivre was arguably the first person to make this connection explicit. It was to prove fundamental to the development of both probability theory and statistics.

Defining probability

A major conceptual problem in probability theory was to define probability. Even simple examples – where everyone knew the answer – presented logical difficulties. If we toss a coin, then in the long run, we expect equal numbers of heads and tails, and the probability of

What probability did for them

In 1710 John Arbuthnot presented a paper to the Royal Society in which he used probability theory as evidence for the existence of God. He analysed the annual number of christenings for male and female children for the period 1629–1710, and found that there are slightly more boys than girls. Moreover, the figure was pretty much the same in every year. This fact was already well known, but Arbuthnot proceeded to calculate the probability of the proportion being constant. His result was very small, 2^{-82}. He then pointed out that if the same effect occurs in all countries, and at all times throughout history, then the chances are even smaller, and concluded that divine providence, not chance, must be responsible.

On the other hand, in 1872 Francis Galton used probabilities to estimate the efficacy of prayer by noting that prayers were said every day, by huge numbers of people, for the health of the royal family. He collected data and tabulated the 'mean age attained by males of various classes who had survived their 30th year, from 1758 to 1843,' adding that 'deaths by accident are excluded.' These classes were eminent men, royalty, clergy, lawyers, doctors, aristocrats, gentry, tradesmen, naval officers, literature and science, army officers, and practitioners of the fine arts. He found that 'The sovereigns are literally the shortest lived of all who have the advantage of affluence. The prayer has therefore no efficacy, unless the very questionable hypothesis be raised, that the conditions of royal life may naturally be yet more fatal, and that their influence is partly, though incompletely, neutralized by the effects of public prayers.'

each is ½. More precisely, this is the probability provided the coin is fair. A biased coin might always turn up heads. But what does 'fair' mean? Presumably, that heads and tails are equally likely. But the phrase 'equally likely' refers to probabilities. The logic seems circular. In order to define probability, we need to know what probability is.

The way out of this impasse is one that goes back to Euclid, and was brought to perfection by the algebraists of the late 19th and early 20th centuries. Axiomatize. Stop worrying about what probabilities are. Write down the properties that you want probabilities to possess, and consider those to be axioms. Deduce everything else from them.

The question was: what are the right axioms? When probabilities refer to finite sets of events, this question has a relatively easy answer. But applications of probability theory often involved choices from potentially infinite sets of possibilities. If you measure the angle between two stars, say, then in principle that can be any real number between 0° and 180°. There are infinitely many real numbers. If you throw a dart at a board, in such a way that in the long run it has the same chance of hitting any point on the board, then the probability of hitting a given region should be the area of that region, divided by the total area of the dartboard. But there are infinitely many points on a dartboard, and infinitely many regions.

These difficulties caused all sorts of problems, and all sorts of paradoxes. They were finally resolved by a new idea from analysis, the concept of a measure.

Analysts working on the theory of integration found it necessary to go beyond Newton, and define increasingly sophisticated notions of what constitutes an integrable function, and what its integral is. After a series of attempts by various mathematicians, Henri Lebesgue managed to define a very general type of integral, now called the Lebesgue integral, with many pleasant and useful analytic properties.

The key to his definition was *Lebesgue measure*, which is a way to assign a concept of length to very complicated subsets of the real line. Suppose that the set consists of non-overlapping intervals of lengths 1, ½, ¼, ⅛ and so on. These numbers form a convergent series, with sum 2. So Lebesgue insisted that this set has measure 2. Lebesgue's concept has a new feature: it is *countably additive*. If you put together an infinite collection of non-overlapping sets, and if this collection is countable in Cantor's sense, with cardinal $\aleph_0$, then the measure of the whole set is the sum of the infinite series formed by the measures of the individual sets.

In many ways the idea of a measure was more important than the integral to which it led. In particular, probability is a measure. This property was made explicit in the 1930s by Andrei Kolmogorov, who laid down axioms for probabilities. More precisely, he defined a *probability space*. This comprises a set X, a collection B of subsets of X called *events* and a measure m on B. The axioms state that m is a measure, and that $m(X) = 1$ (that is, the probability that *something* happens is always 1.) The collection B is also required to have some set-theoretic properties that allow it to support a measure.

For a die, the set X consists of the numbers 1–6, and the set B contains every subset of X. The measure of any set Y in B is the number of members of Y, divided by 6. This measure is consistent with the intuitive idea that each face of the die has probability $1/6$ of turning up. But the use of a measure requires us to consider not just faces, but sets of faces. The probability associated with some such set Y is the probability that some face in Y will turn up. Intuitively, this is the size of Y divided by 6.

With this simple idea, Kolmogorov resolved several centuries of often heated controversy, and created a rigorous theory of probability.

Statistical data

The applied arm of probability theory is statistics, which uses probabilities to analyse real world data. It arose from 18th century astronomy, when observational errors had to be taken into account. Empirically and theoretically, such errors are distributed according to the error function or normal distribution, often called the bell curve because of its shape. Here the error is measured horizontally, with zero error in the middle, and the height of the curve represents the probability of an error of given size. Small errors are fairly likely, whereas large ones are very improbable.

In 1835 Adolphe Quetelet advocated using the bell curve to model social data – births, deaths, divorces, crime and suicide. He discovered that although such events are unpredictable for individuals, they have statistical patterns when observed for an entire population. He personified this idea in terms of the 'average man', a fictitious individual who was average in every respect. To Quetelet, the average man was not just a mathematical concept: it was the goal of social justice.

From about 1880 the social sciences began to make extensive use of statistical ideas, especially the bell curve, as a substitute for experiments. In 1865 Francis Galton made a study of human heredity. How does the height of a child relate to that of its parents? What about weight, or intellectual ability? He adopted Quetelet's bell curve, but saw it as a method for separating distinct populations, not as a moral imperative. If certain data showed two peaks, rather

What probability does for us

A very important use of probability theory occurs in medical trials of new drugs. These trials collect data on the effects of the drugs – do they seem to cure some disorder, or do they have unwanted adverse effects? Whatever the figures seem to indicate, the big question here is whether the data are statistically significant. That is, do the data result from a genuine effect of the drug, or are they the result of pure chance? This problem is solved using statistical methods known as *hypothesis testing*. These methods compare the data with a statistical model, and estimate the probability of the results arising by chance. If, say, that probability is less than 0.01, then with probability 0.99 the data are *not* due to chance. That is, the effect is significant at the 99% level. Such methods make it possible to determine, with considerable confidence, which treatments are effective, or which produce adverse effects and should not be used.

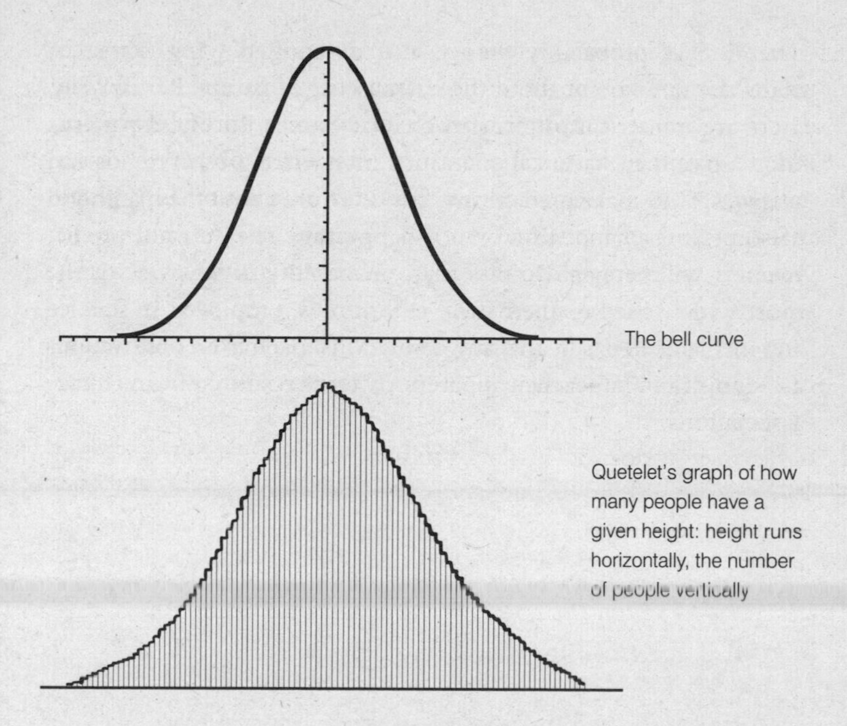

The bell curve

Quetelet's graph of how many people have a given height: height runs horizontally, the number of people vertically

than the single peak of the bell curve, then that population must be composed of two distinct sub-populations, each following its own bell curve. By 1877 Galton's investigations were leading him to invent regression analysis, a way of relating one data set to another to find the most probable relationship.

Another key figure was Francis Ysidro Edgeworth. Edgeworth lacked Galton's vision, but he was a far better technician, and he put Galton's ideas on a sound mathematical basis. A third was Karl Pearson, who developed the mathematics considerably. But Pearson's most effective role was that of salesman: he convinced the outside world that statistics was a useful subject.

Newton and his successors demonstrated that mathematics can be a very effective way to understand the regularities of nature. The invention of probability theory, and its applied wing, statistics, made the same point about the *irregularities* of nature. Remarkably, there are numerical patterns in chance events. But these patterns show up only in statistical quantities such as long-term trends and averages. They make predictions, but these are about the likelihood of some event happening or not happening. They do not predict when it will happen. Despite that, probability is now one of the most widely used mathematical techniques, employed in science and medicine to ensure that any deductions made from observations are significant, rather than apparent patterns resulting from chance associations.

CHAPTER 19

Number Crunching

Calculating machines and computational mathematics

Mathematicians have always dreamed of building machines to reduce the drudgery of routine calculations. The less time you spend calculating, the more time you can spend thinking. From prehistoric times sticks and pebbles were used as an aid to counting, and piles of pebbles eventually led to the abacus, in which beads slide along rods to represent the digits of numbers. Especially as perfected by the Japanese, the abacus could carry out basic arithmetic rapidly and accurately in the hands of an expert. Around 1950 a Japanese abacus out-performed a mechanical hand-calculator.

A dream comes true?

By the 21st century, the advent of electronic computers and the widespread availability of integrated circuits (chips) had given machines a massive advantage. They were far faster than the human brain or a mechanical device – billions or trillions of arithmetical operations every second are now commonplace. The fastest as I write,

IBM's Blue Gene/L, can perform one quadrillion calculations (floating-point operations) per second. Today's computers also have vast memory, storing the equivalent of hundreds of books ready for almost instant recall. Colour graphics have reached a pinnacle of quality.

The rise of the computer

Earlier machines were more modest, but they still saved a lot of time and effort. The first development after the abacus was probably Napier's bones, or Napier's rods, a system of marked rods that Napier invented before he came up with logarithms. Essentially they were the universal components of traditional long multiplication. The rods could be used in place of pencil and paper, saving time writing numerals, but they mimicked hand calculations.

In 1642 Pascal invented the first genuinely mechanical calculator, the Arithmetic Machine, to help his father with his accounts. It could perform addition and subtraction, but not multiplication or division. It had eight rotating dials, so it effectively worked with eight-digit numbers. In the succeeding decade, Pascal built fifty similar machines, most of which are preserved in museums to this day.

In 1671 Leibniz designed a machine for multiplication, and built one in 1694, remarking that 'It is unworthy of excellent men to lose hours like slaves in the labour of calculation, which could be safely relegated to anyone else if machines were used.' He named his machine the Staffelwalze (step reckoner). His main idea was widely used by his successors.

One of the most ambitious proposals for a calculating machine was made by Charles Babbage. In 1812 he said he 'was sitting in the rooms of the Analytical Society, at Cambridge, my head leaning forward on the table in a kind of dreamy mood, with a table of logarithms lying open before me. Another member, coming into the room, and seeing me half asleep, called out, "Well, Babbage, what are you dreaming about?" to which I replied "I am thinking that all these tables" (pointing to the logarithms) "might be calculated by machinery".'

Babbage pursued this dream for the rest of his life, constructing a prototype called the difference engine. He sought government funding for more elaborate machines. His most ambitious project, the analytical engine, was in effect a programmable mechanical computer. None of these machines was built, although various components were made. A modern reconstruction of the difference engine is in the Science Museum in London – and it works. Augusta Ada King, Countess of Lovelace, contributed to Babbage's work by developing some of the first computer programs ever written.

Augusta Ada King

1815 – 52

Augusta Ada was the daughter of the poet Lord Byron and Anne Milbanke. Her parents separated a month after her birth and she never saw her father again. The child showed mathematical ability, and unusually Lady Byron thought that mathematics was good training for the mind and encouraged her daughter to study it. In 1833 Ada met Charles Babbage at a party, and soon after she was shown his prototype difference engine, finding it fascinating and rapidly understanding how it worked. She became Countess of Lovelace when her husband William was created an Earl in 1838.

In her 1843 translation of Luigi Menabrea's *Notions sur la Machine Analytique de Charles Babbage* she added what are in effect sample programs of her own devising. She wrote that 'The distinctive characteristic of the Analytical Engine ... is the introduction into it of the principle which Jacquard devised for regulating, by means of punched cards, the most complicated patterns in the fabrication of brocaded stuffs ... We may say most aptly that the Analytical Engine weaves algebraical patterns just as the Jacquard loom weaves flowers and leaves.'

At the age of 36 she developed uterine cancer, and died after a lengthy period of severe pain, bled to death by her physicians.

The first mass-produced calculator, the Arithmometer, was manufactured by Thomas de Colmar in 1820. It employed a stepped drum mechanism and was still in production in 1920. The next major step was the pin wheel mechanism of the Swedish inventor Willgodt T. Odhner. His calculator set the pattern for dozens, if not hundreds of similar machines, made by a variety of manufacturers. The motive power was supplied by the operator, who turned a handle to revolve a series of discs on which the digits 0–9 were displayed. With practice, complicated calculations could be carried out at high speed. The scientific and engineering calculations for the Manhattan project of the Second World War, to make the first atomic bomb, were performed using such machines by a squad of 'calculators' – mainly young women. The advent of cheap powerful electronic computers in the 1980s made mechanical calculators obsolete, but their use in business and scientific computing was widespread until that time.

Calculating machines contribute more than just simple arithmetic, because many scientific calculations can be implemented numerically as lengthy series of arithmetical operations. One of the earliest numerical methods, which solves equations to arbitrarily high precision, was invented by Newton, and is appropriately known as *Newton's method*. It solves an equation $f(x) = 0$ by calculating a series of successive approximations to a solution, each improving on the previous one but based on it. From some initial guess, x_1, improved approximations $x_2, x_3, \ldots, x_n, x_{n+1}$, are derived using the formula

$$x_{n+1} = x_n - \frac{f(x_n)}{f'(x_n)}$$

where f' is the derivative of f. The method is based on the geometry of the curve $y = f(x)$ near the solution. The point x_{n+1} is where the tangent to the curve at x_n crosses the x-axis. As the diagram shows, this is closer to x than the original point.

A second important application of numerical methods is to differential equations. Suppose we wish to solve the differential equation

$$\frac{dx}{dt} = f(x)$$

given that $x = x_0$ at time $t = 0$. The simplest method, due to Euler, is to approximate $\frac{dx}{dt}$ by $\frac{x(t+\varepsilon) - x(t)}{\varepsilon}$, where ε is very small. Then an approximation to the differential equation takes the form

$$x(t + \varepsilon) = x(t) + \varepsilon f(x(t))$$

Starting with $x(0) = x_0$, we successively deduce the values of $f(\varepsilon)$, $f(2\varepsilon)$, $f(3\varepsilon)$ and, in general, $f(n\varepsilon)$ for any integer $n > 0$. A typical value for ε might be 10^{-6}, say. A million iterations of the formula tells us $x(1)$, another million leads to $x(2)$ and so on. With today's computers a million calculations are trivial, and this becomes an entirely practical method.

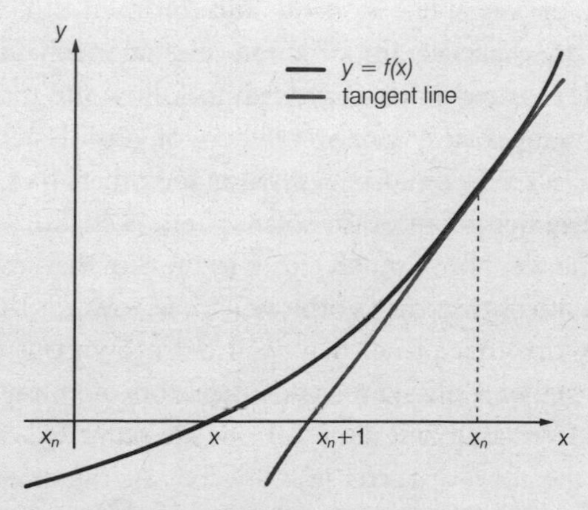

Newton's method for solving an equation numerically

However, the Euler method is too simple-minded to be fully satisfactory, and numerous improvements have been devised. The best known are a whole class of Runge–Kutta methods, named after the German mathematicians Carl Runge and Martin Kutta, who invented their first such method in 1901. One of these, the so-called *fourth order* Runge–Kutta method, is very widely used in engineering, science and theoretical mathematics.

The needs of modern nonlinear dynamics have spawned several sophisticated methods that avoid accumulating errors over long periods of time by preserving certain structure associated with the exact solution. For example, in a mechanical system without friction, the total energy is conserved. It is possible to set up the numerical method so that at each step energy is conserved *exactly*. This procedure avoids the possibility that the computed solution will slowly drift away from the exact one, like a pendulum that is slowly coming to rest as it loses energy.

More sophisticated still are symplectic integrators, which solve mechanical systems of differential equations by explicitly and exactly preserving the symplectic structure of Hamilton's equations, which is a curious but hugely important kind of geometry tailored to the two types of variable, position and momentum. Symplectic integrators are especially important in celestial mechanics, where – for example – astronomers may wish to follow the movements of the planets in the solar system over billions of years. Using symplectic integrators, Jack Wisdom, Jacques Laskar and others have shown that the long-term behaviour of the solar system is chaotic, that Uranus and Neptune were once much closer to the Sun than they are now, and that eventually Mercury's orbit will move towards that of Venus, so that one or other planet may well be thrown out of the solar system altogether. Only symplectic integrators offer any confidence that results over such long time periods are accurate.

Computers need mathematics

As well as using computers to help mathematics, we can use mathematics to help computers. In fact, mathematical principles were important in all early computer designs, either as proof of concept or as key aspects of the design.

All digital computers in use today work with binary notation, in which numbers are represented as strings of just two digits: 0 and 1. The main advantage of binary is that it corresponds to switching: 0 is off and 1 is on. Or 0 is no voltage and 1 is 5 volts, or whatever standard is employed in circuit design. The symbols 0 and 1 can also be interpreted within mathematical logic, as *truth values*: 0 means false and 1 means true. So computers can perform logical calculations as well as arithmetical ones. Indeed, the logical operations are more basic, and the arithmetical operations can be viewed as sequences of logical operations. Boole's algebraic approach to the mathematics of 0 and 1, in *The Laws of Thought*, provides an effective formalism for the logic of computer calculations. Internet search engines carry out Boolean searches, that is, they search for items defined by some combination of logical criteria, such as 'containing the word "cat" but not containing "dog"'.

What numerical analysis did for them

Newton not only had to sort out patterns in nature: he had to develop effective methods of calculation. He made extensive use of power series to represent functions, because he could differentiate or integrate such series term by term. He also used them to calculate values of functions, an early numerical method still in use today. One page of his manuscripts, dating to 1665, shows a numerical calculation of the area under a hyperbola, which we now recognize as the logarithmic function. He added together the terms of an infinite series, working to an astonishing 55 decimal places.

Algorithms

Mathematics has aided computer science, but in return computer science has motivated some fascinating new mathematics. The notion of an *algorithm* – a systematic procedure for solving a problem – is one. (The name comes from the Arabian algebraist al-Khwarizmi.) An especially interesting question is: how does the running time of an algorithm depend on the size of the input data?

For example, Euclid's algorithm for finding the highest common factor of two whole numbers m and n, with $m \leq n$, is as follows:

- Divide n by m to get remainder r.
- If $r = 0$ then the highest common factor is m: STOP.
- If $r > 0$ then replace n by m and m by r and go back to the start.

It can be shown that if n has d decimal digits (a measure of the size of the input data to the algorithm) then the algorithm stops after at most $5d$ steps. That means, for instance, that if we are given two 1000-digit numbers, we can compute their highest common factor in at most 5000 steps – which takes a split second on a modern computer.

The Euclidean algorithm has linear running time: the length of the computation is proportional to the size (in digits) of the input data. More generally, an algorithm has polynomial running time, or is of class P, if its running time is proportional to some fixed power (such as the square or cube) of the size of the input data. In contrast, all known algorithms for finding the prime factors of a number have exponential running time – some fixed constant raised to the power of the size of the input data. This is what makes the RSA cryptosystem (conjecturally) secure.

Roughly speaking, algorithms with polynomial running time lead to practical computations on today's computers, whereas algorithms with exponential running time do not – so the corresponding computations cannot be performed in practice, even for quite small sizes of initial data. This distinction is a rule of thumb: a polynomial algorithm might involve such a large power that it is impractical, and some algorithms with running time worse than polynomial still turn out to be useful.

The main theoretical difficulty now arises. Given a specific algorithm, it is (fairly) easy to work out how its running time depends on the size of the input data and to determine whether it is class P or not. However, it is extraordinarily difficult to decide whether a more efficient algorithm might exist to solve the same problem more rapidly. So, although we know that many problems can be solved by an algorithm in class P, we have no idea whether any sensible problem is not-P.

Sensible here has a technical meaning. Some problems *must* be not-P, simply because *outputting the answer* requires non-P running time. For example list all possible ways to arrange n symbols in order. To rule out such obviously non-P problems, another concept is needed: the class NP of *non-deterministic* polynomial algorithms. An algorithm is NP if any guess at an answer can be checked in a time proportional to some fixed power of the size of input data. For example, given a guess at a prime factor of a large number, it can quickly be checked by a single division sum.

A problem in class P is automatically NP. Many important problems, for which P algorithms are not known, are also known to be NP. And now we come to the deepest and most difficult unsolved problem in this area, the solution of which will win a million-dollar prize from the Clay Mathematics Institute. Are P and NP the same? The most plausible answer is no, because P = NP means that a lot of apparently very hard computations are actually easy – there exists some short cut which we've not yet thought of.

The P = NP? problem is made more difficult by a fascinating phenomenon, called NP-completeness. Many NP problems are such that if they are actually class P, then *every* NP problem is class P as well. Such a problem is said to be NP-*complete*. If any particular NP-complete problem can be proved to be P, then P = NP. On the other hand, if any particular NP-complete problem can be proved to be not-P, then P is not the same as NP. One NP-complete problem that attracted attention recently is associated with the computer game Minesweeper. A more mathematical one is the Boolean Satisfiability Problem: given a statement in mathematical logic, can it ever be true for some assignment of truth values (true or false) for its variables?

Numerical analysis

Mathematics involves much more than calculations, but calculations are an unavoidable accompaniment to more conceptual investigations. From the earliest times, mathematicians have sought mechanical aids to free them from the drudgery of calculation, and to improve the likelihood of accurate results. The mathematicians of the past would have envied our access to electronic computers, and marvelled at their speed and accuracy.

Calculating machines have done more for mathematics than just act as servants. Their design and function have posed new theoretical questions for mathematicians to answer. These questions range from justifying approximate numerical methods for solving equations, to deep issues in the foundations of computation itself.

At the start of the 21st century, mathematicians now have access to powerful software, making it possible not just to perform numerical calculations on computers, but to perform algebraic and analytic ones. These tools have opened up new areas, helped to solve long-standing problems, and freed up time for conceptual thinking. Mathematics has become much richer as a result, and it has also become applicable to many more practical problems. Euler had the conceptual tools to study fluid flow round complicated shapes, and even though aircraft had not been invented, there are plenty of interesting questions about ships in water. But he did not have any practical methods to implement those techniques.

What numerical analysis does for us

Numerical analysis plays a central role in the design of modern aircraft. Not so long ago, engineers worked out how air would flow past the wings and fuselage of an aircraft using wind-tunnels. They would place a model of the aircraft in the tunnel, force air past it with a system of fans and observe the flow patterns. Equations such as those of Navier and Stokes provided various theoretical insights, but it was impossible to solve them for real aircraft because of their complicated shape.

Today's computers are so powerful, and numerical methods for solving PDEs on computers have become so effective, that in many cases the wind-tunnel approach has been discarded in favour of a numerical wind-tunnel – that is, a computer model of the aircraft. The Navier–Stokes equations are so accurate that they can safely be used in this way. The advantage of the computer approach is that any desired feature of the airflow can be analysed and visualized.

A new development, not touched on above, is the use of computers as an aid to proof. Several important theorems, proved in recent years, rely on massive but routine calculations, carried out by computer. It has been argued that computer-assisted proofs change the fundamental nature of proof, by removing the requirement that the proof can be verified by a human mind. This assertion is controversial, but even if it is true, the result of the change is to make mathematics into an even more powerful aid to human thought.

CHAPTER 20

Chaos and Complexity

Irregularities have patterns too

By the middle of the 20th century, mathematics was undergoing a rapid phase of growth, stimulated by its widespread applications and powerful new methods. A comprehensive history of the modern period of mathematics would occupy at least as much space as a treatment of everything that led up to that period. The best we can manage is a few representative samples, to demonstrate that originality and creativity in mathematics are alive and well. One such topic, which achieved public prominence in the 1970s and 1980s, is chaos theory, the media's name for nonlinear dynamics. This topic evolved naturally from traditional models using calculus. Another is complex systems, which employs less orthodox ways of thinking, and is stimulating new mathematics as well as new science.

Chaos

Before the 1960s, the word chaos had only one meaning: formless disorder. But since that time, fundamental discoveries in science and mathematics have endowed it with a second, more subtle, meaning – one that combines aspects of disorder with aspects of form. Newton's *Mathematical Principles of Natural Philosophy* had reduced the system of the world to differential equations, and these are *deterministic*. That is, once the initial state of the system is known, its future is determined uniquely for all time. Newton's vision is that of a clockwork universe, set in motion by the hand of the creator but thereafter pursuing a single inevitable path. It is a vision that seems to leave no scope for free will, and this may well be one of the early sources of the belief that science is cold and inhuman. It is also a vision that has served us well, giving us radio, television, radar, mobile phones, commercial aircraft, communications satellites, man-made fibres, plastics and computers.

The growth of scientific determinism was also accompanied by a vague but deep-seated belief in the conservation of complexity. This is the assumption that simple causes must produce simple effects, implying that complex effects must have complex causes. This belief causes us to look at a complex object or system, and wonder where the complexity comes from. Where, for example, did the complexity of life come from, given that it must have originated on a lifeless planet? It seldom occurs to us that complexity might appear of its own accord, but that is what the latest mathematical techniques indicate.

A single solution?

The determinacy of the laws of physics follows from a simple mathematical fact: there is at most one solution to a differential equation with given initial conditions. In Douglas Adams's *The Hitchhiker's Guide to the Galaxy* the supercomputer Deep Thought embarked on a five million year quest for the answer to the great question of life, the universe and everything, and famously derived the answer, 42. This incident is a parody of a famous statement in which Laplace summed up the mathematical view of determinism:

> 'An intellect which at any given moment knew all the forces that animate nature and the mutual positions of the beings that comprise it, if this intellect were vast enough to submit its data to analysis, it could condense into a single formula the movement of the greatest bodies of the universe and that of the lightest atom: for such an intellect nothing could be uncertain, and the future just like the past would be present before its eyes.'

He then brought his readers down to earth with a bump, by adding:

> 'Human mind offers a feeble sketch of this intelligence in the perfection which it has been able to give to astronomy'.

Ironically, it was in celestial mechanics, that most evidently deterministic part of physics, that Laplacian determinism would meet a sticky end. In 1886 King Oscar II of Sweden (who also ruled Norway) offered a prize for solving the problem of the stability of the solar system. Would our own little corner of the clockwork universe keep ticking forever, or might a planet crash into the Sun or escape into interstellar space? Remarkably, the physical laws of conservation of energy and momentum do not forbid either eventuality – but could the detailed dynamics of the solar system shed further light?

Poincaré was determined to win the prize, and he warmed up by taking a close look at a simpler problem, a system of three celestial bodies. The equations for three bodies are not much worse than those for two, and have much the same general form. But Poincaré's three-body warm-up turned out to be surprisingly hard, and he discovered something disturbing. The solutions of those equations were totally different from those of the two-body case. In fact, the solutions were so complicated that they could not be written down as a mathematical formula. Worse, he could understand enough of the geometry – more precisely, the topology – of the solutions to prove, beyond any shadow of doubt, that the motions represented by those solutions could sometimes be highly disordered and irregular. 'One is struck,' Poincaré wrote, 'with the complexity of this figure that I am not even attempting to draw. Nothing can give us a better idea of the complexity of the three-body problem.' This complexity is now seen as a classic example of chaos.

His entry won King Oscar II's prize, even though it did not fully solve the problem posed. Some 60 years later, it triggered a revolution in how we view the universe and its relation to mathematics.

In 1926–7 the Dutch engineer Balthazar van der Pol constructed an electronic circuit to simulate a mathematical model of the heart, and discovered that under certain conditions the resulting oscillation is not periodic, like a normal heartbeat, but irregular. His work was given a solid mathematical basis during the Second World War by John Littlewood and Mary Cartwright, in a study that originated in the electronics of radar. It took more than 40 years for the wider significance of their work to become apparent.

Nonlinear dynamics

In the early 1960s the American mathematician Stephen Smale ushered in the modern era of dynamical systems theory by asking for a complete classification of the typical types of behaviour of electronic circuits. Originally expecting the answer to be combinations of periodic motions, he quickly realized that much more complicated behaviour is possible. In particular he developed Poincaré's discovery of complex motion in the restricted three-body problem, simplifying the geometry to yield a system known as 'Smale's horsehoe'. He proved that the horsehoe system, although deterministic, has some random features. Other examples of such phenomena were developed by the American and Russian schools of dynamics, with especially notable contributions by Oleksandr

Sharkovskii and Vladimir Arnold, and a general theory began to emerge. The term 'chaos' was introduced by James Yorke and Tien-Yien Li in 1975, in a brief paper that simplified one of the results of the Russian school: Sharkovskii's Theorem of 1964, which described a curious pattern in the periodic solutions of a discrete dynamical system – one in which time runs in integer steps instead of being continuous..

Meanwhile, chaotic systems were appearing sporadically in the applied literature – again, largely unappreciated by the wider scientific community. The best known of these was introduced by the meteorologist Edward Lorenz in 1963. Lorenz set out to model

Poincaré's Blunder

June Barrow-Green, delving into the archives of the Mittag-Leffler Institute in Stockholm, recently discovered an embarrassing tale that had previously been kept quiet. The work that had won Poincaré the prize contained a serious mistake. Far from discovering chaos, as had been supposed, he had claimed to prove that it did not occur. The original submission proved that all motions in the three-body problem are regular and well-behaved.

After the award of the prize, Poincaré belatedly spotted an error, and quickly discovered that it demolished his proof completely. But the prize-winning memoir had already been published as an issue of the Institute's journal. The journal was withdrawn, and Poincaré paid for a complete reprint, which included his discovery of homoclinic tangles and what we now call chaos. This cost him significantly more than the money he had won with his flawed memoir. Almost all of the copies of the incorrect version were successfully reclaimed and destroyed, but one, preserved in the archives of the Institute, slipped through the net.

atmospheric convection, approximating the very complex equations for this phenomenon by much simpler equations with three variables. Solving them numerically on a computer, he discovered that the solution oscillated in an irregular, almost random manner. He also discovered that if the same equations are solved using initial values of the variables that differ only very slightly from each other, then the differences become amplified until the new solution differs completely from the original one. His description of this phenomenon in subsequent lectures led to the currently popular term, butterfly effect, in which the flapping of a butterfly's wings leads, a month later, to a hurricane on the far side of the globe.

This weird scenario is a genuine one, but in a rather subtle sense. Suppose you could run the world's weather twice: once with the butterfly and once without. Then you would indeed find major differences, quite possibly including a hurricane on one run but no hurricane on the other. Exactly this effect arises in computer simulations of the equations normally used to predict the weather, and the effect causes big problems in weather forecasting. But it would be a mistake to conclude that the butterfly *caused* the hurricane. In the real world, weather is influenced not by one butterfly but by the statistical features of trillions of butterflies and other tiny disturbances. Collectively, these have a definite influence on where and when hurricanes form, and where they subsequently go.

Using topological methods, Smale, Arnold and their coworkers proved that the bizarre solutions observed by Poincaré were the inevitable consequence of *strange attractors* in the equations. A strange attractor is a complex motion that the system inevitably homes in on. It can be visualized as a shape in the state-space formed by the variables that describe the system. The Lorenz attractor, which describes Lorenz's equations in this manner, looks a bit like the Lone Ranger's mask, but each apparent surface has infinitely many layers.

The structure of attractors explains a curious feature of chaotic systems: they can be predicted in the short term (unlike, say, rolls of a die) but not in the long term. Why cannot several short-term predictions be strung together to create a long-term prediction? Because the accuracy with which we can describe a chaotic system deteriorates over time, at an ever-growing rate, so there is a prediction horizon beyond which we cannot penetrate. Nevertheless, the system remains on the same strange attractor – but its path over the attractor changes significantly.

This modifies our view of the butterfly effect. All the butterflies can do is push the weather around on the same strange attractor – so it always looks like perfectly plausible weather. It's just a bit different from what it would have been without all those butterflies.

Mary Lucy Cartwright

1900–98

Mary Cartwright graduated from Oxford University in 1923, one of only five women studying mathematics in the university. After a short period as a teacher, she took a doctorate at Cambridge, nominally under Godfrey Hardy but actually under Edward Titchmarsh because Hardy was in Princeton. Her thesis topic was complex analysis. In 1934 she was appointed assistant lecturer at Cambridge, and in 1936 she was made director of studies at Girton College.

In 1938, in collaboration with John Littlewood, she undertook research for the Department of Scientific and Industrial Research on differential equations related to radar. They discovered that these equations had highly complicated solutions, an early anticipation of the phenomenon of chaos. For this work she became the first woman mathematician to be elected a Fellow of the Royal Society, in 1947. In 1948 she was made Mistress of Girton, and from 1959 to 1968 she was a reader at the University of Cambridge. She received many honours, and was made Dame Commander of the British Empire in 1969.

David Ruelle and Floris Takens quickly found a potential application of strange attractors in physics: the baffling problem of turbulent flow in a fluid. The standard equations for fluid flow, called the Navier–Stokes equations, are partial differential equations, and as such they are deterministic. A common type of fluid flow, laminar flow, is smooth and regular, just what you would expect from a deterministic theory. But another type, turbulent flow, is frothy and irregular, almost random. Previous theories either claimed that turbulence was an extremely complicated combination of patterns that individually were very simple and regular, or that the Navier–Stokes equations broke down in the turbulent regime. But Ruelle and Takens had a third theory. They suggested that turbulence is a physical instance of a strange attractor.

Initially this theory was received with some scepticism, but we now know that it was correct in spirit, even if some of the details were rather questionable. Other successful applications followed, and the word chaos was enlisted as a convenient name for all such behaviour.

Theoretical monsters

A second theme now enters our tale. Between 1870 and 1930 a diverse collection of maverick mathematicians invented a series of bizarre shapes the sole purpose of which was to show up the limitations of classical analysis. During the early development of calculus, mathematicians had assumed that any continuously varying quantity must possess a well-defined rate of change almost everywhere. For example, an object that is moving through space continuously has a well-defined speed, except at relatively few instants when its speed changes abruptly. However, in 1872 Weierstrass showed that this long-standing assumption is wrong. An object can move in a continuous fashion, but in such an irregular manner that – in effect – its speed is changing abruptly at every moment of time. This means that it doesn't actually *have* a sensible speed at all.

Other contributions to this strange zoo of anomalies included a curve that fills an entire region of space (one was found by Peano in 1890, another by Hilbert in 1891), a curve that crosses itself at every point (discovered by Waclaw Sierpinski in 1915) and a curve of infinite length that encloses a finite area. This last example of geometric weirdness, invented by Helge von Koch in 1906, is the *snowflake curve*, and its construction goes like this: Begin with an equilateral triangle, and add triangular promontories to the middle of each side to create a six-pointed star. Then add smaller promontories to the middle of the star's twelve sides, and keep going forever. Because of its sixfold symmetry, the results look like an intricate snowflake. Real snowflakes grow by other rules, but that's a different story.

The mainstream of mathematics promptly denounced these oddities as 'pathological' and a 'gallery of monsters', but as the years went by several embarrassing fiascos emphasized the need for care, and the mavericks' viewpoint gained ground. The logic behind analysis is so subtle that leaping to plausible conclusions is

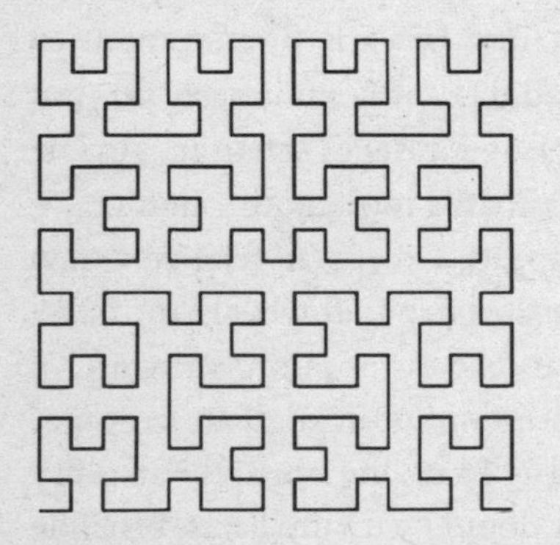

Stages in the construction of Hilbert's space-filling curve and the Sierpinski gasket

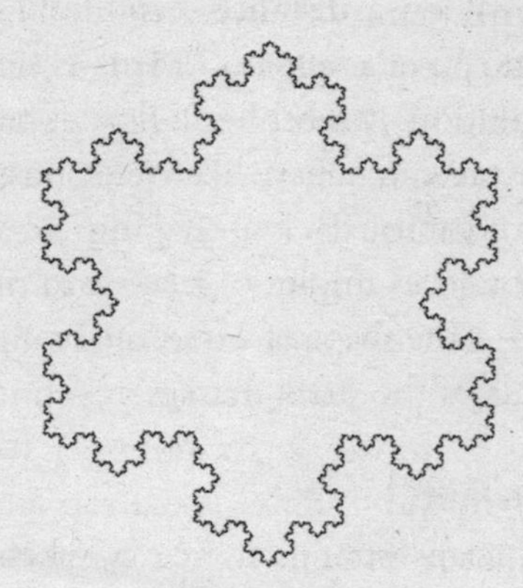

The snowflake curve

dangerous: monsters warn us about what can go wrong. So, by the turn of the century, mathematicians had become comfortable with the newfangled goods in the mavericks' curiosity shop – they kept the theory straight without having any serious impact on applications. Indeed by 1900 Hilbert could refer to the whole area as a paradise without causing ructions.

In the 1960s, against all expectations, the gallery of theoretical monsters was given an unexpected boost in the direction of applied science. Benoit Mandelbrot realized that these monstrous curves are clues to a far-reaching theory of irregularities in nature. He renamed them *fractals*. Until then, science had been happy to stick to traditional geometric forms like rectangles and spheres, but Mandelbrot insisted that this approach was far too restrictive. The natural world is littered with complex and irregular structures – coastlines, mountains, clouds, trees, glaciers, river systems, ocean waves, craters, cauliflowers – about which traditional geometry remains mute. A new geometry of nature is needed.

Today, scientists have absorbed fractals into their normal ways of thinking, just as their predecessors did at the end of the 19th century with those maverick mathematical monstrosities. The second half of Lewis Fry Richardson's 1926 paper 'Atmospheric diffusion shown on a distance–neighbour graph' bears the title 'Does the wind have a velocity?'. This is now seen as an entirely reasonable question. Atmospheric flow is turbulent, turbulence is fractal and fractals can behave like Weierstrass's monstrous function – moving continuously but having no well defined speed. Mandelbrot found examples of fractals in many areas both in and outside science – the shape of a tree, the branching pattern of a river, the movements of the stock market.

Chaos everywhere!

The mathematicians' strange attractors, when viewed geometrically, turned out to be fractals, and the two strands of thought became intertwined in what is now popularly known as chaos theory.

Chaos can be found in virtually every area of science. Jack Wisdom and Jacques Laskar have found that the dynamics of the solar system is chaotic. We know all the equations, masses and speeds that are required to predict the future motion forever, but there is a prediction horizon of around ten million years because of dynamical chaos. So if you want to know what side of the Sun Pluto will be in AD 10,000,000 – forget it. These astronomers have also shown that the Moon's tides stabilize the Earth against influences that would otherwise lead to chaotic motion, causing rapid shifts of climate from warm periods to ice ages and back again; so chaos theory demonstrates that without the Moon, the Earth would be a pretty unpleasant place to live.

Chaos arises in nearly all mathematical models of biological populations, and recent experiments (letting beetles breed under controlled conditions) indicate that it arises in real biological populations as well. Ecosystems do not normally settle down to some kind of static balance of nature: instead they wander around on strange attractors, usually looking fairly similar, but always changing. The failure to understand the subtle dynamics of ecosystems is one reason why the world's fisheries are close to disaster.

Complexity

From chaos, we turn to complexity. Many of the problems facing today's science are extremely complicated. To manage a coral reef, a forest or a fishery it is necessary to understand a highly complex ecosystem, in which apparently harmless changes can trigger unexpected problems. The real world is so complicated, and can be so difficult to measure, that conventional modelling methods are hard to set up and even harder to verify. In response to these challenges, a growing number of scientists have come to believe that fundamental changes are needed in the way we model our world.

In the early 1980s George Cowan, formerly head of research at Los Alamos, decided that one way forward lay in the newly developed theories of nonlinear dynamics. Here small causes can create huge effects, rigid rules can lead to anarchy and the whole often has capabilities that do not exist, even in rudimentary form, in its components. In general terms, these are exactly the features observed in the real world. But does the similarity run deeper – deep enough to provide genuine understanding?

Cowan conceived the idea of a new research institute devoted to the interdisciplinary applications and development of nonlinear dynamics. He was joined by Murray Gell-Mann, the Nobel-winning particle physicist, and in 1984 they created what was then called the Rio Grande Institute. Today it is the Santa Fe Institute, an international centre for the study of complex systems. Complexity theory has contributed novel mathematical methods and viewpoints, exploiting computers to create digital models of nature. It exploits the power of the computer to analyse those models and deduce tantalizing features of complex systems. And it uses nonlinear dynamics and other areas of mathematics to understand what the computers reveal.

Cellular automaton

In one type of new mathematical model, known as a *cellular automaton*, such things as trees, birds and squirrels are incarnated as tiny coloured squares. They compete with their neighbours in a mathematical computer game. The simplicity is deceptive – these games lie at the cutting edge of modern science.

Cellular automata came to prominence in the 1950s, when John von Neumann was trying to understand life's ability to copy itself. Stanislaw Ulam suggested using a system introduced by the computer pioneer Konrad Zuse in the 1940s. Imagine a universe composed of a large grid of squares, called *cells*, like a giant chessboard. At any moment, a given square can exist in some state. This chessboard universe is equipped with its own laws of nature, describing how each cell's state must change as time clicks on to the next instant. It is useful to represent that state by colours. Then the rules would be statements like: 'If a cell is red and has two blue cells next to it, it must turn yellow.' Any such system is called a cellular automaton – cellular because of the grid, automaton because it blindly obeys whatever rules are listed.

To model the most fundamental feature of living creatures, von Neumann created a configuration of cells that could replicate – make copies of itself. It had 200,000 cells and employed 29 different colours to carry around a coded description of itself. This description could be copied blindly, and used as a blueprint for building further configurations of the same kind. Von Neumann did not publish his work until 1966, by which time Crick and Watson had discovered the structure of DNA and it had become clear how life really does perform its replication trick. Cellular automata were ignored for another 30 years.

By the 1980s, however, there was a growing interest in systems composed of large numbers of simple parts, which interact to produce a complicated whole. Traditionally, the best way to model a system mathematically is to include as much detail as possible: the closer the model is to the real thing, the better. But this high-detail approach fails

What Nonlinear dynamics did for them

Until nonlinear dynamics became a major issue in scientific modelling, its role was mainly theoretical. The most profound work was that of Poincaré on the three-body problem in celestial mechanics. This predicted the existence of highly complex orbits, but gave little idea of what they looked like. The main point of the work was to demonstrate that simple equations may not have simple solutions – that complexity is not conserved, but can have simpler origins.

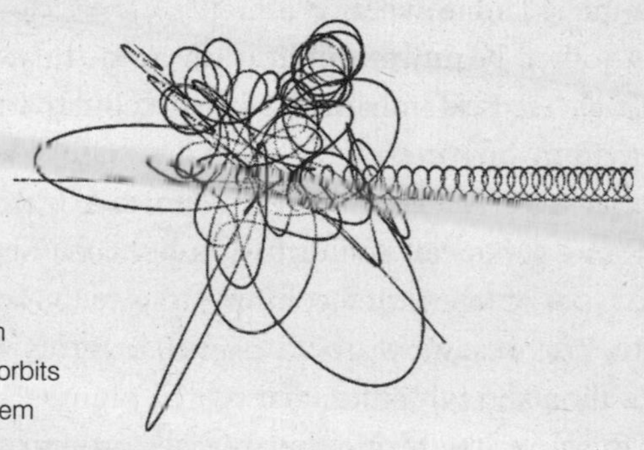

Modern computers can calculate complicated orbits in the three-body problem

for very complex systems. Suppose, for instance, that you want to understand the growth of a population of rabbits. You don't need to model the length of the rabbits' fur, how long their ears are or how their immune systems work. You need only a few basic facts about each rabbit: how old it is, what sex it is, whether it is pregnant. Then you can focus your computer resources on what really matters.

For this kind of system, cellular automata are very effective. They make it possible to ignore unnecessary detail about the individual components, and instead to focus on how these components interrelate. This turns out to be an excellent way to work out which factors are important, and to uncover general insights into why complex systems do what they do.

Geology and biology

A complex system that defies analysis by traditional modelling techniques is the formation of river basins and deltas. Peter Burrough has used cellular automata to explain why these natural features adopt the shapes that they do. The automaton models the interactions between water, land and sediment. The results explain how different rates of soil erosion affect the shapes of rivers, and how rivers carry soil away, important questions for river engineering and management. The ideas are also of interest to oil companies, because oil and gas are often found in geological strata that were originally laid down as sediment.

Another beautiful application of cellular automata occurs in biology. Hans Meinhardt has used cellular automata to model the formation of patterns on animals, from seashells to zebras. Key factors are concentrations of chemicals. Interactions are reactions within a given cell and diffusion between neighbouring cells. The two types of interaction combine to give the actual rules for the next state. The results provide useful insights into the patterns of activation and inhibition that switch pigment-making genes on and off dynamically during the animal's growth.

Stuart Kauffman has applied a variety of complexity-theoretic techniques to delve into another major puzzle in biology: the development of organic form. The growth and development of an organism must involve a great deal of dynamics, and it cannot be just a matter of translating into organic form the information held in DNA. A promising way forward is to formulate development as the dynamics of a complex nonlinear system.

Cellular automata have now come full circle and given us a new perspective on the origins of life. Von Neumann's self-replicating automaton is enormously special, carefully tailored to make copies of one highly complex initial configuration. Is this typical of self-replicating automata, or can we obtain replication without starting from a very special configuration? In 1993 Hui-Hsien Chou and

James Reggia developed a cellular automaton with 29 states for which a randomly chosen initial state, or primordial soup, leads to self-replicating structures more than 98 per cent of the time. In this automaton, self-replicating entities are a virtual certainty.

Complex systems support the view that on a lifeless planet with sufficiently complex chemistry, life is likely to arise spontaneously and to organize itself into ever more complex and sophisticated forms. What remains to be understood is what kinds of rule lead to the spontaneous emergence of self-replicating configurations in our own universe – in short, what kind of physical laws make this first crucial step towards life not only possible, but inevitable.

How mathematics was created

The story of mathematics is long and convoluted. As well as making remarkable breakthroughs, the pioneers of mathematics headed off down blind alleys, sometimes for centuries. But this is the way of the pioneer. If it is obvious where to go next, anyone can do it. And so, over some four millennia, the elaborate, elegant structure that we call mathematics came into being. It arose in fits and starts, with wild bursts of activity followed by periods of stagnation; the centre of activity moved around the globe following the rise and fall of human culture. Sometimes it grew according to the practical needs of that culture; sometimes the subject took off in its own direction, as its practitioners played what seemed to everyone else to be mere intellectual games. And surprisingly often, those games eventually paid off in the real world, by stimulating the development of new techniques, new points of view and new understanding.

Mathematics has not stopped. New applications demand new mathematics, and mathematicians are responding. Biology, especially, poses new challenges to mathematical modelling and understanding. The internal requirements of mathematics continue to stimulate new ideas, new theories. Many important conjectures remain unsolved, but mathematicians are working on them.

What Nonlinear dynamics does for us

It might seem that chaos has no practical applications, being irregular, unpredictable and highly sensitive to small disturbances. However, because chaos is based on deterministic laws, it turns out to be useful precisely because of these features.

One of the potentially most important applications is chaotic control. Around 1950 the mathematician John von Neumann suggested that the instability of weather might one day be turned to advantage, because it implies that a large *desired* effect can be generated by a very small disturbance. In 1979 Edward Belbruno realized that this effect could be used in astronautics to move spacecraft through large distances with very little expenditure of fuel. However, the resulting orbits take a long time – two years from the Earth to the Moon, for instance – and NASA lost interest in the idea.

In 1990 Japan launched a small Moon probe, Hagoromo, which separated from a larger probe Hiten that stayed in Earth orbit. But the radio on Hagoromo failed, leaving Hiten with no role to play. Japan wanted to salvage something from the mission, but Hiten had only 10% of the fuel required to get it to the Moon using a conventional orbit. An engineer on the project remembered Belbruno's idea, and asked him to help. Within ten months Hiten was on its way to the Moon and beyond, seeking trapped particles of interstellar dust – with half of its remaining fuel unused. The technique has been used repeatedly since this first success, in particular for the Genesis probe to sample the solar wind, and ESA's SMARTONE mission.

The technique applies on Earth as well as in space. In 1990 Celso Grebogi, Edward Ott and James Yorke published a general theoretical scheme to exploit the butterfly effect in the control of chaotic systems. The method has been used to synchronize a bank of lasers; to control heartbeat irregularities, opening up the possibility of an intelligent pacemaker; to control electrical waves in the brain, which might help to suppress epileptic attacks; and to smooth the motion of a turbulent fluid, which could in future make aircraft more fuel-efficient.

Throughout its lengthy history, mathematics has taken its inspiration from these two sources – the real world and the world of human imagination. Which is most important? Neither. What matters is the combination. The historical method makes it plain that mathematics draws its power, and its beauty, from both. The time of the ancient Greeks is often seen as a historical Golden Age, as logic, mathematics, and philosophy were brought to bear on the human condition. But the advances made by the Greeks are just part of an ongoing story. Mathematics has never been so active, it has never been so diverse, and it has never been so vital to our society.

Welcome to the Golden Age of mathematics.

Further Reading

Books and articles

E. Belbruno, *Fly Me to the Moon*, Princeton University Press, Princeton 2007.

E.T. Bell, *Men of Mathematics* (2 vols.), Pelican, Harmondsworth 1953.

E.T. Bell, *The Development of Mathematics* (reprint), Dover, New York 2000.

R. Bourgne and J.-P. Azra, *Écrits et Mémoires Mathématiques d'Évariste Galois*, Gauthier-Villars, Paris 1962.

C.B. Boyer, *A History of Mathematics*, Wiley, New York 1968.

W.K. Bühler, *Gauss: a Biographical Study*, Springer, Berlin 1981.

J. Cardan, *The Book of My Life* (translated by Jean Stoner), Dent, London 1931.

G. Cardano, *The Great Art or the Rules of Algebra* (translated T. Richard Witmer), MIT Press, Cambridge, MA 1968.

J. Coolidge, *The Mathematics of Great Amateurs*, Dover, New York 1963.

T. Dantzig, *Number – the Language of Science* (ed. J. Mazur), Pi Press, New York 2005.

Euclid, *The Thirteen Books of Euclid's Elements* (3 vols, translated by Sir Thomas L. Heath), Dover, New York 1956.

J. Fauvel and J. Gray, *The History of Mathematics – a Reader*, Macmillan Education, Basingstoke 1987.

D.H. Fowler, *The Mathematics of Plato's Academy*, Clarendon Press, Oxford 1987.

C.F. Gauss, *Disquisitiones Arithmeticae*, Leipzig 1801 (translated by A.A. Clarke), Yale University Press, New Haven, CT 1965.

A. Hyman, *Charles Babbage*, Oxford University Press, Oxford 1984.

G.G. Joseph, *The Crest of the Peacock – non-European Roots of Mathematics*, Penguin, Harmondsworth 2000.

V.J. Katz, *A History of Mathematics* (2nd edn), Addison-Wesley, Reading, MA 1998.

M. Kline, *Mathematical Thought from Ancient to Modern Times*, Oxford University Press, Oxford 1972.

A.H. Koblitz, *A Convergence of Lives – Sofia Kovalevskaia*, Birkhäuser, Boston 1983.

N. Koblitz, *A Course in Number Theory and Cryptography* (2nd edn.), Springer, New York 1994.

M. Livio, *The Equation That Couldn't Be Solved*, Simon & Schuster, New York 2005.

M. Livio, *The Golden Ratio*, Broadway, New York 2002.

E. Maior, *e – the Story of a Number*, Princeton University Press, Princeton 1994.

E. Maior, *Trigonometric Delights*, Princeton University Press, Princeton 1998.

D. McHale, *George Boole*, Boole Press, Dublin 1985.

O. Neugebauer, *A History of Ancient Mathematical Astronomy* (3 vols.) Springer, New York 1975.

O. Ore, *Niels Hendrik Abel: Mathematician Extraordinary*, University of Minnesota Press, Minneapolis 1957.

C. Reid, *Hilbert*, Springer, New York 1970.

T. Rothman, 'The short life of Évariste Galois', *Scientific American* (April 1982) 112–120. Collected in T. Rothman, *A Physicist on Madison Avenue*, Princeton University Press 1991.

D. Sobel, *Longitude* (10th anniversary edn), HarperPerennial, New York 2005.

I. Stewart, *Does God Play Dice? – The New Mathematics of Chaos*, (2nd edition) Penguin, Harmondsworth 1997.

I. Stewart, *Why Beauty is Truth*, Basic Books, New York 2007.

S.M. Stigler, *The History of Statistics*, Harvard University Press, Cambridge, MA 1986.

B.L. van der Waerden, *A History of Algebra*, Springer-Verlag, New York 1994.

D. Welsh, *Codes and Cryptography*, Oxford University Press, Oxford 1988.

Internet

Most topics can be located easily using a search engine. Three very good general sites are:

The MacTutor History of Mathematics archive:
http://www-groups.dcs.st-and.ac.uk/~history/index.html

Wolfram MathWorld, a compendium of information on mathematical topics:
http://mathworld.wolfram.com

Wikipedia, the free online encyclopaedia:
http://en.wikipedia.org/wiki/Main_Page

Index